Crystallographic and Modeling Methods in Molecular Design

Charles E. Bugg   Steven E. Ealick
Editors

Crystallographic and Modeling Methods in Molecular Design

With 106 Figures

Springer-Verlag
New York Berlin Heidelberg
London Paris Tokyo Hong Kong

Charles E. Bugg
Department of Biochemistry
University of Alabama
Birmingham, AL 35294
USA

Steven E. Ealick
Department of Biochemistry
University of Alabama
Birmingham, AL 35294
USA

Library of Congress Cataloging-in-Publication Data
Crystallographic and modeling methods in molecular design/Charles E.
Bugg, Steven E. Ealick.
    p. cm.
    Papers presented at the Crystallographic and Modeling Methods in
Molecular Design Symposium in Gulf Shores, Alabama, April 30–May 3,
1989, and sponsored by the Department of Biochemistry, Univ. of
Alabama at Birmingham.
    ISBN-13:978-1-4612-7987-7
    1. Drugs—Design—Congresses.   2. Drugs—Mathematical models—
Congresses.   3. Crystallography—Congresses.   4. Molecular
structure—Congresses.   I. Bugg, Charles E.   II. Ealick, Steven E.
III. Crystallographic and Modeling Methods in Molecular Design
Symposium (1989: Gulf Shores, Ala.)   IV. University of Alabama at
Birmingham. Dept. of Biochemistry.
RS420.C79   1990   89-26201
615'.19—dc20

Printed on acid-free paper.

© 1990 Springer-Verlag New York, Inc.
Softcover reprint of the hardcover 1st edition 1990

Camera-ready copy supplied by the authors.

9 8 7 6 5 4 3 2 1

ISBN-13:978-1-4612-7987-7   e-ISBN-13:978-1-4612-3374-9
DOI:10.1007/978-1-4612-3374-9

Preface

This book contains the papers that were presented at the "Crystallographic and Modeling Methods in Molecular Design Symposium" in Gulf Shores, Alabama, April 30 to May 3, 1989.

During the past few years, there has been a burst of activity in this area, especially related to drug design and protein engineering projects. The purpose of the symposium and this book is to provide an up-to-date review of the most recent experimental and theoretical approaches that are being used for molecular design. The book covers several recent examples of approaches for using crystallography in conjunction with forefront modeling methods for guiding the development of enzyme inhibitors and of peptides and proteins with modified biological and physical properties. In addition, this book contains discussions of new approaches for combining crystallographic data and advanced computational techniques for aiding in the design of enzyme inhibitors and other compounds that bind to selected biological targets. This book is therefore of interest not only to molecular biologists and biochemists, but is stimulating reading for anyone involved in structural biology, pharmaceutical chemistry, enzymology, protein engineering, and biotechnology.

The meeting was the third in a series of symposia initiated and sponsored by the Department of Biochemistry, University of Alabama at Birmingham. We gratefully acknowledge the generous financial support of the following institutions and companies:

- Department of Biochemistry, University of Alabama at Birmingham, Birmingham, Alabama;
- Center for Macromolecular Crystallography, University of Alabama at Birmingham, Birmingham, Alabama;
- Office of Commercial Programs, National Aeronautics and Space Administration (Grant NAGW-813), Washington, D.C.;
- International Union of Crystallography, Chester, England;

- BioCryst, Inc., Birmingham, Alabama;
- Burroughs-Wellcome Company, Raleigh, North Carolina;
- Ciba-Geigy Corporation, Summit, New Jersey;
- Dow-Elanco Chemical Company, Midland, Michigan;
- E.I. du Pont de Nemours and Company, Wilmington, Delaware;
- Eastman Kodak Company, Rochester, New York;
- Eli Lilly and Company, Indianapolis, Indiana;
- Merck Sharp and Dohme Research Laboratories, Rahway, New Jersey;
- Schering Corporation, Bloomfield, New Jersey;
- Smith Kline and French Laboratories, Philadelphia, Pennsylvania; and
- The Upjohn Company, Kalamazoo, Michigan.

We would like to express our appreciation to Mary Ann Lynch and Cookie Woodruff for the time and hard work that they devoted to making this symposium a success. We thank Penny Mann for help in assembling the manuscripts and for editorial assistance with this book.

Birmingham, Alabama   CHARLES E. BUGG
                                                   STEVEN E. EALICK

Contents

Preface v

Contributors and Participants ix

Inhibitor Binding to Thymidylate Synthase Is Mediated by
Different Structural Determinants than Those That Promote
Tight Binding to Dihydrofolate Reductase 1
DAVID A. MATTHEWS, CHERYL A. JANSON, and WARD W. SMITH

Crystallographic and Pharmacological Studies of Antiviral Agents
Against Human Rhinovirus 9
THOMAS J. SMITH, JOHN BADGER, MARCIA KREMER, MARCOS
OLIVEIRA, MICHAEL G. ROSSMANN, MARK A. MCKINLAY, GUY D.
DIANA, DANIEL C. PAVEAR, FRANK J. DUTKO, ROLAND R.
RUECKERT, BEVERLY HEINZ, and DEBORAH SHEPARD

Structural Studies of Elastase-Inhibitor Complexes with
β-Lactams 29
MANUEL A. NAVIA, BRIAN M. MCKEEVER, and JAMES P.
SPRINGER

Design of Purine Nucleoside Phosphorylase Inhibitors Using
X-Ray Crystallography 43
STEVEN E. EALICK, Y.S. BABU, S.V.L. NARAYANA, WILLIAM J.
COOK, and CHARLES E. BUGG

The Potential Role of Solvation in the Dihydrofolate Reductase
Species Selectivity of Trimethoprim 56
LEE F. KUYPER

Crystallographic and Genetic Approaches Toward the Design of
Proteins of Enhanced Thermostability 80
J.A. WOZNIAK, X.-J. ZHANG, K. WILSON, L.H. WEAVER,
D.E. TRONRUD, P.E. PJURA, H. NICHOLSON, M. MATSUMURA,

viii

M. Karpusas, R. Jacobson, R. Faber, S. Dao-pin, J.A. Bell,
T. Alber, and Brian W. Matthews

Stability of Folded Conformations by Computer Simulation:
Methods and Some Applications 95
Jan Hermans, R.-H. Yun, and Amil G. Anderson

The Use of Molecular Dynamics and Free Energy Perturbation
Approaches in Simulating the Properties of Macromolecules and
Their Binding to Ligands 114
Peter A. Kollman

Molecular Recognition of DNA Minor Groove Binding Drugs 123
Andrew H.-J. Wang and Mai-kun Teng

Structural and Computational Studies of Anticonvulsants: A
Search for Correlation Between Molecular Systematics and
Activity 151
Penelope W. Codding, N.E. Duke, L.J. Aha, L.Y. Palmer,
D.K. McClurg, and M.B. Szkaradzinska

Crystallography and Molecular Mechanics in Designing Drugs
with Unknown Receptor Structure 161
David J. Duchamp

Molecular Modeling with Substructure Libraries Derived from
Known Protein Structures 175
Barry C. Finzel, S. Kimatian, D.H. Ohlendorf,
J.J. Wendoloski, M. Levitt, and F. Ray Salemme

O: A Macromolecule Modeling Environment 189
T. Alwyn Jones, Marc Bergdoll, and Morten Kjeldgaard

Inhibitor Design from Known Structures 200
Renee L. DesJarlais, Brian Shoichet, George Seibel, and Irwin
D. Kuntz, Jr.

The Cambridge Structural Database in Molecular Modeling:
Systematic Conformational Analysis from Crystallographic Data .. 211
Frank H. Allen and Michael J. Doyle

Preferred Interaction Patterns from Crystallographic Databases ... 229
R. Scott Rowland, Frank H. Allen, W. Michael Carson, and
Charles E. Bugg

Aladdin: A Real Tool for Structure-Based Drug Design 254
Yvonne C. Martin

Index 265

Contributors and Participants

L.J. AHA, Departments of Chemistry, Pharmacology, and Therapeutics, University of Calgary, Calgary, Alberta, Canada, T2N 1N4.

T. ALBER, Institute of Molecular Biology and Departments of Physics, Chemistry, and Biology, University of Oregon, Eugene, OR 97403, USA.

FRANK H. ALLEN, University Chemistry Laboratory, University of Cambridge, Cambridge CB2 1EW, England.

AMIL G. ANDERSON, Department of Biochemistry, School of Medicine, University of North Carolina, Chapel Hill, NC 27599-7260, USA.

SHELLY ARMSTRONG, University of Alabama at Birmingham, Birmingham, AL 35294, USA.

BERNADETTE ARNOUX, C.N.R.S., 91198 Gif sur Yvette, France.

Y.S. BABU, Center for Macromolecular Crystallography, University of Alabama at Birmingham, Birmingham, AL 35294, USA.

JOHN BADGER, Department of Biological Sciences, Purdue University, West Lafayette, IN 47907, USA.

J.A. BELL, Institute of Molecular Biology and Departments of Physics, Chemistry, and Biology, University of Oregon, Eugene, OR 97403, USA.

MARC BERGDOLL, Department of Molecular Biology, University of Uppsala, S-75124. Uppsala, Sweden. *Present address:* Lab. de Cristallographie Biologique, IBMC, 67084 Strasbourg Cedex, France.

JAY BERTRAND, Department of Chemistry, Georgia Institute of Technology, Atlanta, GA 30332, USA.

TESFAYE BIFTU, Merck Sharp and Dohme Research Laboratories, P.O. Box 2000, Rahway, NJ 07065, USA.

PAT BOSSART, University of Alabama at Birmingham, Birmingham, AL 35294, USA.

BARBARA BRANDHUBER, Synergen, 1885 33rd Street, Boulder, CO 80301, USA.

CHRISTIE G. BROUILLETTE, Department of Gerontology and Geriatric Medicine, University of Alabama at Birmingham, Birmingham, AL 35294, USA.

WAYNE J. BROUILLETTE, Department of Chemistry, University of Alabama at Birmingham, Birmingham, AL 35294, USA.

CHARLES E. BUGG, Center for Macromolecular Crystallography, University of Alabama at Birmingham, Birmingham, AL 35294, USA.

ROBERT CAMERON, University of New Orleans, 2000 Lakeshore Drive, New Orleans, LA 70122, USA.

W. MICHAEL CARSON, Center for Macromolecular Crystallography, University of Alabama at Birmingham, Birmingham, AL 35294, USA.

PENELOPE W. CODDING, Departments of Chemistry, Pharmacology, and Therapeutics, University of Calgary, Calgary, Alberta, Canada T2N 1N4.

WILLIAM J. COOK, Center for Macromolecular Crystallography, University of Alabama at Birmingham, Birmingham, AL 35294, USA.

S. DAO-PIN, Institute of Molecular Biology and Departments of Physics, Chemistry, and Biology, University of Oregon, Eugene, OR 97403, USA.

RENEE L. DESJARLAIS, Department of Pharmaceutical Chemistry, School of Pharmacy, University of California, San Francisco, CA 94143, USA.

GUY D. DIANA, Sterling Research Group, 81 Columbia Turnpike, Rensselaer, NY 12144, USA.

DABNEY DIXON, Department of Chemistry, Georgia State University, University Plaza, Atlanta, GA 30303, USA.

MICHAEL J. DOYLE, Crystallographic Data Centre, University Chemical Laboratory, Lensfield Road, Cambridge, CB2 1EW, England.

DAVID J. DUCHAMP, Physical and Analytical Chemistry Research, The Upjohn Company, 301 Henrietta Street, Kalamazoo, MI 49001, USA.

N.E. DUKE, Departments of Chemistry, Pharmacology, and Therapeutics, University of Calgary, Calgary, Alberta, Canada, T2N 1N4.

FRANK J. DUTKO, Sterling Research Group, 81 Columbia Turnpike, Rensselaer, NY 12144, USA.

STEVEN E. EALICK, Center for Macromolecular Crystallography, University of Alabama at Birmingham, Birmingham, AL 35294, USA.

MARK ERION, Ciby-Geigy Corporation, 556 Morris Avenue, Summit, NJ 07901, USA.

R. FABER, Institute of Molecular Biology and Departments of Physics, Chemistry, and Biology, University of Oregon, Eugene, OR 97403, USA.

KRZYSZTOF A. FIDELIS, Department of Chemistry, The University of Oklahoma, Norman, OK 73019, USA.

BARRY C. FINZEL, Central Research and Development Department, Du Pont Experimental Station ES228/320, Wilmington, DE 19880-0228, USA.

JIM FOUNTAIN, Marshall Space Flight Center, Code PS05, Marshall Space Flight Center, AL 35812, USA.

ALFRED FRENCH, USDA-SRRC, P.O. Box 19687, New Orleans, LA 70179, USA.

LARRY GARTLAND, Department of Microbiology, University of Alabama at Birmingham, Birmingham, AL 35294, USA.

PIET GROS, Chemische Laboratoria Der Ryksuniversiteit Groningen, Nijenborgh 16, 9747 AG Groningen, The Netherlands.

WAYNE C. GUIDA, Ciby-Geigy Corporation, 556 Morris Avenue, Summit, NJ 07901, USA.

BEVERLY HEINZ, Institute for Molecular Virology, University of Wisconsin, Madison, WI 53706, USA.

ALICE HELLER, Center for Macromolecular Crystallography, University of Alabama at Birmingham, Birmingham, AL 35294, USA.

JAN HERMANS, Department of Biochemistry, School of Medicine, University of North Carolina at Chapel Hill, Chapel Hill, NC 27514, USA.

DOUGLAS M. Ho, Department of Chemistry, University of Cincinnati, Cincinnati, OH 45221, USA.

RUTH LEH-YEH HSU, The Ohio State University, Columbus, OH 43210, USA.

ERIC HUNTER, Department of Microbiology, University of Alabama at Birmingham, Birmingham, AL 35294, USA.

R. JACOBSON, Institute of Molecular Biology and Departments of Physics, Chemistry, and Biology, University of Oregon, Eugene, OR 97403, USA.

CHERYL A. JANSON, Agouron Pharmaceuticals, 11025 North Torrey Pines Road, La Jolla, CA 92037, USA.

GREG JENKINS, Teledyne-Brown Engineering, 300 Sparkman Drive, Huntsville, AL 35807, USA.

T. ALWYN JONES, Department of Molecular Biology, University of Uppsala, S-75124 Uppsala, Sweden.

M. KARPUSAS, Institute of Molecular Biology and Departments of Physics, Chemistry, and Biology, University of Oregon, Eugene, OR 97403, USA.

JUNG JA P. KIM, Medical College of Wisconsin, Milwaukee, WI 53201, USA.

S. KIMATIAN, Central Research and Development Department, Du Pont Experimental Station ES228/320, Wilmington, DE 19880-0228, USA.

MORTEN KJELDGAARD, Department of Chemistry, University of Aarhus, DK-8000 Aarhus, Denmark.

PETER A. KOLLMAN, Department of Pharmaceutical Chemistry, School of Pharmacy, University of California, San Francisco, CA 94143, USA.

MARCIA KREMER, Department of Biological Sciences, Purdue University, West Lafayette, IN 47907, USA.

RAMU KRISHNA, Department of Biochemistry, University of Alabama at Birmingham, Birmingham, AL 35294, USA.

IRWIN D. KUNTZ, JR., Department of Pharmaceutical Chemistry, School of Pharmacy, University of California, San Francisco, CA 94143, USA.

LEE F. KUYPER, Department of Experimental Therapy, Burroughs-Wellcome Company, 3030 Cornwallis Road, Research Triangle Park, NC 27709, USA.

ANTONI LACZKOWSKI, Texas A & M University, College Station, TX 77840, USA.

MIKE LAWRENCE, Division of Biotechnology, CSIRO, 343 Royal Parade, Parkville, 3052 Victoria, Australia.

M. LEVITT, Department of Cell Biology, Stanford University, Stanford, CA 94305, USA.

MARIANNA M. LONG, Center for Macromolecular Crystallography, University of Alabama at Birmingham, Birmingham, AL 35294, USA.

THOMAS J. LUKAS, Howard Hughes Medical Institute, Vanderbilt University, Nashville, TN 37232, USA.

MING LUO, Center for Macromolecular Crystallography, University of Alabama at Birmingham, Birmingham, AL 35294, USA.

JOSEPH A. MADDRY, Southern Research Institute, P.O. Box 55305, Birmingham, AL 35255, USA.

YVONNE C. MARTIN, Abbott Laboratories, Department 47E, AP9, Abbott Park, IL 60064, USA.

M. MATSUMURA, Institute of Molecular Biology and Departments of Physics, Chemistry, and Biology, University of Oregon, Eugene, OR 97403, USA.

BRIAN W. MATTHEWS, Institute of Molecular Biology, University of Oregon, Eugene, OR 97403, USA.

DAVID A. MATTHEWS, Agouron Pharmaceuticals, 505 Coast Boulevard South, La Jolla, CA 92037, USA.

D.K. McCLURG, Departments of Chemistry, Pharmacology, and Therapeutics, University of Calgary, Calgary, Alberta, Canada, T2N 1N4.

BRIAN M. McKEEVER, Merck Sharp and Dohme Research Laboratories, P.O. Box 2000, Rahway, NJ 07065-0900, USA.

MARK A. McKINLAY, Sterling Research Group, 81 Columbia Turnpike, Rensselaer, NY 12144, USA.

GRETCHEN MEINKE, Department of Chemistry, Georgia Institute of Technology, Atlanta, GA 30332, USA.

EDGAR MEYER, Texas A & M University, College Station, TX 77843, USA.

VLADIMIR MINIC, Department of Biochemistry, University of Alabama at Birmingham, Birmingham, AL 35294, USA.

S. LAKSHMI NARASHIMHAN, Department of Chemistry, Georgia Institute of Technology, Atlanta, GA 30332, USA.

S.V.L. NARAYANA, Center for Macromolecular Crystallography, University of Alabama at Birmingham, Birmingham, AL 35294, USA.

ROBERT J. NAUMANN, Marshall Space Flight Center, Code ES-71, Marshall Space Flight Center, AL 35812, USA.

MANUEL A. NAVIA, Merck Sharp and Dohme Research Laboratories, P.O. Box 2000, Rahway, NJ 07065-0900, USA.

H. NICHOLSON, Institute of Molecular Biology and Departments of Physics, Chemistry, and Biology, University of Oregon, Eugene, OR 97403, USA.

D.H. OHLENDORF, Central Research and Development Department, Du Pont Experimental Station ES228/320, Wilmington, DE 19880-0228, USA.

MARCOS OLIVEIRA, Department of Biological Sciences, Purdue University, West Lafayette, IN 47907, USA.

L.Y. PALMER, Departments of Chemistry, Pharmacology, and Therapeutics, University of Calgary, Calgary, Alberta, Canada, T2N 1N4.

HANS E. PARGE, RIXC-MB-5, 10666 North Torrey Pines Road, La Jolla, CA 92037, USA.

DANIEL C. PAVEAR, Sterling Research Group, 81 Columbia Turnpike, Rensselaer, NY 12144, USA.

BERIT F. PEDERSEN, Institute of Pharmacy, University of Oslo, Blindern, 0316 Oslo 3, Norway.

SUSAN R. PHILLIPS, Department of Chemistry and Biochemistry, Georgia Institute of Technology, Atlanta, GA 30332, USA.

P.E. PJURA, Institute of Molecular Biology and Departments of Physics, Chemistry, and Biology, University of Oregon, Eugene, OR 97403, USA.

JAMES C. POWERS, Georgia Institute of Technology, Atlanta, GA 30332, USA.

R. RADHAKRISHNAN, Department of Biochemistry and Biophysics, Texas A & M University, College Station, TX 77843, USA.

JAMES RAFTERY, University of Manchester, Manchester M13 9PL, England.

PINO RAUCCI, Department of Biochemistry, University of Alabama at Birmingham, Birmingham, AL 35294, USA.

ROBERT C. REYNOLDS, Southern Research Institute, P.O. Box 55035, Birmingham, AL 35255, USA.

WILLIAM H. RILEY, The Dow-Elanco Chemical Company, 1701 Building, Midland, MI 48640, USA.

MICHAEL G. ROSSMAN, Department of Biological Sciences, Purdue University, West Lafayette, IN 47907, USA.

R. SCOTT ROWLAND, Center for Macromolecular Crystallography, University of Alabama at Birmingham, Birmingham, AL 35294, USA.

ROLAND R. RUECKERT, Institute for Molecular Virology, University of Wisconsin, Madison, WI 53706, USA.

MICHAL SABAT, Department of Chemistry, Northwestern University, Evanston, IL 60208, USA.

F. RAY SALEMME, Central Research and Development Department, Du Pont Experimental Station, P.O. Box 80228, Wilmington, DE 19880-0228, USA.

GEORGE SEIBEL, Department of Pharmaceutical Chemistry, School of Pharmacy, University of California, San Francisco, CA 94143, USA.

SHOBHA SENADHI, Center for Macromolecular Crystallography, University of Alabama at Birmingham, Birmingham, AL 35294, USA.

VIJAY SENADHI, Center for Macromolecular Crystallography, University of Alabama at Birmingham, Birmingham, AL 35294, USA.

DEBORAH SHEPARD, Institute for Molecular Virology, University of Wisconsin, Madison, WI 53706, USA.

BRIAN SHOICHET, Department of Pharmaceutical Chemistry, School of Pharmacy, University of California, San Francisco, CA 94143, USA.

OLIVER SMART, Biophysics Group, Imperial College, London SW7 2BZ, England.

OLAV SMISTAD, Space Industries, Inc., 711 West Bay Area Boulevard, Houston, TX 77598, USA.

CRAIG SMITH, Center for Macromolecular Crystallography, University of Alabama at Birmingham, Birmingham, AL 35294, USA.

THOMAS J. SMITH, Department of Biological Sciences, Purdue University, West Lafayette, IN 47907, USA.

WARD W. SMITH, Agouron Pharmaceuticals, 11025 North Torrey Pines Road, La Jolla, CA 92037, USA.

JOHN T. SPITZNAGEL, BioCryst, Inc., 1075 13th Street South, Birmingham, AL 35205, USA.

JAMES P. SPRINGER, Merck Sharp and Dohme Research Laboratories, P.O. Box 2000, Rahway, NJ 07065-0900, USA.

JERRY STEWART, Center for Macromolecular Crystallography, University of Alabama at Birmingham, Birmingham, AL 35294, USA.

ROSEMARIE SWANSON, Biochemistry Department, Texas A & M University, College Station, TX 77843-2128, USA.

STANLEY SWANSON, Biochemistry Department, Texas A & M University, College Station, TX 77843-2128, USA.

M.B. SZKARADZINSKA, Departments of Chemistry, Pharmacology, and Therapeutics, University of Calgary, Calgary, Alberta, Canada, T2N 1N4.

MAI-KUN TENG, Department of Physiology and Biophysics, University of Illinois at Urbana-Champaign, Urbana, IL 61801, USA.

D.E. TRONRUD, Institute of Molecular Biology and Departments of Physics, Chemistry, and Biology, University of Oregon, Eugene, OR 97403, USA.

DENNIS J. UNDERWOOD, Merck Sharp and Dohme Research Laboratories, P.O. Box 2000, Rahway, NJ 07065, USA.

JANAKIRAMAN VIJAYALAKSHMI, Biochemistry Department, Texas A & M University, College Station, TX 77843-2128, USA.

DONALD VOET, Department of Chemistry, University of Pennsylvania, Philadelphia, PA 19174, USA.

MARK WALTER, Center for Macromolecular Crystallography, University of Alabama at Birmingham, Birmingham, AL 35294, USA.

ANDREW H.-J. WANG, Department of Physiology and Biophysics, University of Illinois at Urbana-Champaign, Urbana, IL 61801, USA.

L.H. WEAVER, Institute of Molecular Biology and Departments of Physics, Chemistry, and Biology, University of Oregon, Eugene, OR 97403, USA.

PATRICIA C. WEBER, Du Pont Experimental Station, P.O. Box 80228, Wilmington, DE 19880-0228, USA.

ROBERT WELLS, Department of Biochemistry, University of Alabama at Birmingham, Birmingham, AL 35294, USA.

J.J. WENDOLOSKI, Central Research and Development Department, Du Pont Experimental Station ES228/320, Wilmington, DE 19880-0228, USA.

K. WILSON, Institute of Molecular Biology and Departments of Physics, Chemistry, and Biology, University of Oregon, Eugene, OR 97403, USA.

J.A. WOZNIAK, Institute of Molecular Biology and Departments of Physics, Chemistry, and Biology, University of Oregon, Eugene, OR 97403, USA.

JONG-CHANG WU, University of New Orleans, 2000 Lakeshore Drive, New Orleans, LA 70122, USA.

R.-H. YUN, Department of Biochemistry, School of Medicine, University of North Carolina at Chapel Hill, Chapel Hill, NC 27599-7260, USA.

X.-J. ZHANG, Institute of Molecular Biology and Departments of Physics, Chemistry, and Biology, University of Oregon, Eugene, OR 97403, USA.

BAO-GUANG ZHAO, Center for Macromolecular Crystallography, University of Alabama at Birmingham, Birmingham, AL 35294, USA.

HARMON J. ZUCCOLA, Department of Chemistry, Georgia Institute of Technology, Atlanta, GA 30332, USA.

Inhibitor Binding to Thymidylate Synthase Is Mediated by Different Structural Determinants than Those That Promote Tight Binding to Dihydrofolate Reductase

DAVID A. MATTHEWS, CHERYL A. JANSON, and WARD W. SMITH

INTRODUCTION

Thymidylate synthase(TS) catalyzes the conversion of 2'deoxyuridylate(dUMP) to 2'deoxythymidylate(dTMP) utilizing the cofactor 5,10-methylenetetrahydrofolate(CH_2-H_4PteGlu) as both the single carbon donor and reductant. Because the folate cofactor is utilized stoichiometrically rather than catalytically, cells must have a mechanism for regenerating CH_2-H_4PteGlu in order to ensure a continuing supply of dTMP for DNA synthesis. This requirement is met by the so-called thymidylate synthesis cycle (Figure 1) in which dihydrofolate, the other product of the TS catalyzed reaction, is sequentially reduced to tetrahydrofolate by dihydrofolate reductase(DHFR) and finally converted back to the cofactor CH_2-H_4PteGlu by the action of a third enzyme, serine hydroxymethyltransferase. In principal, inhibitors directed at any of these three enzymes could be useful antimetabolites because in shutting down this cycle, the cell is deprived of its only *de novo* source of dTMP.

Of the three enzymes comprising the thymidylate synthesis cycle, DHFR has been the most thoroughly investigated as a target for design of specific inhibitors. The first compounds capable of inducing folate deficiency in animals were reported in 1947 (Franklin et al., 1947; Seeger et al., 1947) more than ten years before it was demonstrated at the molecular level that these agents were acting at DHFR(Osborn et al., 1958). The discovery by Hitchings *et al.*(1948) that certain diaminopyrimidines were acting as antifols and could be engineered to discriminate between enzymes from different species was the first indication that specific antimicrobial agents could be developed based solely upon their selective inhibition of an enzyme common to both parasites and hosts. Some examples of clinically useful DHFR inhibitors are shown in Figure 2. We will have more to say about several of these compounds later when we examine how they bind to their target enzyme.

Inhibition of TS has been achieved with analogs of both the folate and nucleotide substrates. Inhibitors that bind at the folate site have recently attracted renewed attention following reports of potent TS binding by N-l0 substituted 5,8-dideazafolates (Jones et al., 1985). Most notable among the nucleotides is the useful anti-tumor agent 5-fluorodeoxuridine monophosphate (FdUMP), a metabolite of the antipyrimidines 5-fluorouracil and 5-fluorodeoxyuridine.

Much less well studied as a target for inhibitors than either of the other two is the third enzyme in the cycle, serine hydroxymethyltransferase. We shall limit our discussion to include only DHFR and TS.

Structural aspects of inhibitor binding to DHFR have been exhaustively studied and analyzed over the past decade(for a review see Kraut and Matthews, 1987). Only recently have we been able to examine in similar detail how folates and folate-based inhibitors bind

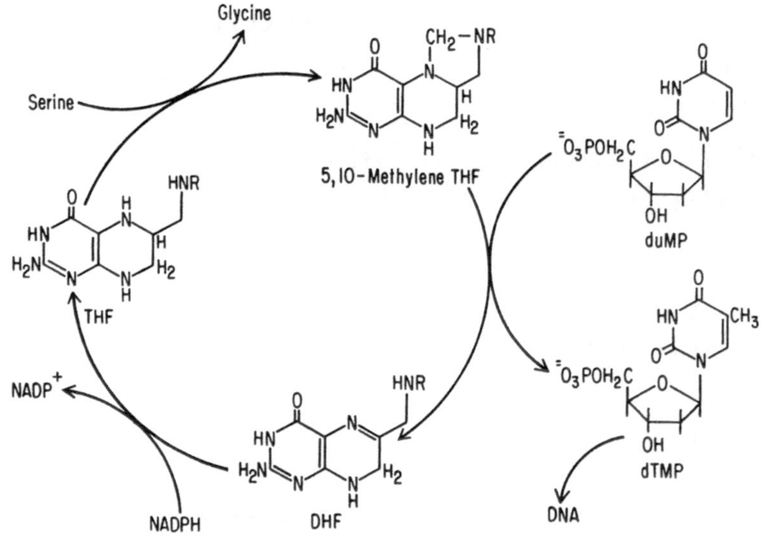

Figure 1. The thymidylate synthesis cycle

to TS. It is perhaps not surprising that since both DHFR and TS possess a folate binding site, powerful inhibitors of both enzymes have been developed based on the folate motif. What is unexpected however is that the portion of a folate-like inhibitor most responsible for tight binding is different in the two cases. In what follows we seek to understand the basis for these differences in terms of the detailed geometrical construction of the two enzyme active sites.

Structure of a Thymidylate Synthase Ternary Complex

We recently reported the refined 2.3Å structure of an *E.coli* TS ternary complex containing FdUMP and 10-propargyl-5,8-dideazafolate (PDDF, Figure 3, Matthews et al., 1989a). The three dimensional structure of *L.casei* TS (3.0Å resolution with isomorphous replacement phases) without bound substrates or inhibitors has been reported by Hardy et al.(1987).

A schematic representation of the polypeptide backbone folding for one of the two chemically identical TS subunits and positioning of the two inhibitors bound at the enzyme's active site is shown in Figure 4. The TS dimer has two widely separated active sites each of which is bounded by three beta strands, two alpha helices and six noncontiguous loop regions including the proteins' carboxy terminus. Each active site cavity is formed by 28 amino acids in all, 26 from one subunit and two from the other. Entrance to the active site is gained via a funnel shaped opening at the back of each protomer on the side opposite the central beta sheet that forms part of the subunit interface. The substrate binding pocket extends from the surface downward into the protein for a distance of about 25Å, terminating near the backside of the large beta sheet. FdUMP binds at the bottom of this cleft where two beta strands separate owing to a kink in the beta sheet caused by an unusual series of three stacked beta bulges (Matthews et al. , 1989b). This splaying apart of two beta strands at the bottom of the nucleotide binding pocket permits close approach of the other subunit to the ligand's 5' phosphate which is securely anchored by charge mediated hydrogen bonds to four conserved arginine sidechains, two from each protomer.

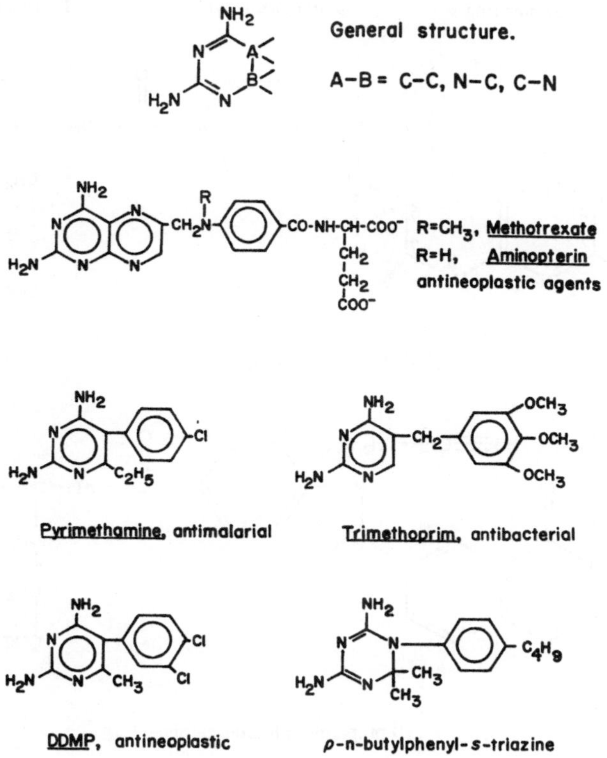

General structure.

A-B = C-C, N-C, C-N

R=CH$_3$, **Methotrexate**
R=H, **Aminopterin**
antineoplastic agents

Pyrimethamine, antimalarial

Trimethoprim, antibacterial

DDMP, antineoplastic

p-n-butylphenyl-s-triazine

Figure 2. Dihydrofolate Reductase Inhibitors

Nucleotide binds to TS in an open conformation covalently linked through C6 to conserved Cys-146. Up-pointing faces of the pyrimidine and ribose rings are fully exposed and unobstructed providing a complementary docking surface for the bicyclic ring of the quinazoline inhibitor. PDDF binds in a partially folded conformation with its p-aminobenzoyl group inclined at a 65° angle to the heterocycle and the inhibitor's glutamate tail exposed at the entrance to the active site cleft. Ternary complex formation induces a conformational change in which the four carboxyl terminal amino acids close down on the distal side of the quinazoline ring capping the active site and sequestering the bound ligands from bulk solvent.

Tight Binding Inhibitors of Dihydrofolate Reductase Contain a 2,4-Diaminoheterocycle

Structural comparison of the DHFR inhibitors grouped in Figure 2 with the enzyme's substrate dihydrofolate suggests that an important determinant of potent inhibition of this enzyme is the presence of a 2,4-diaminoheterocyclic ring comprising the left hand portion of each molecule as drawn in Figure 2. In fact, 2,4-diaminopyrimidine itself is a micromolar inhibitor of chromosomal DHFRs. Substitution patterns in the right hand portion of these molecules can either further enhance binding compared to the 2,4-diaminoheterocycle alone

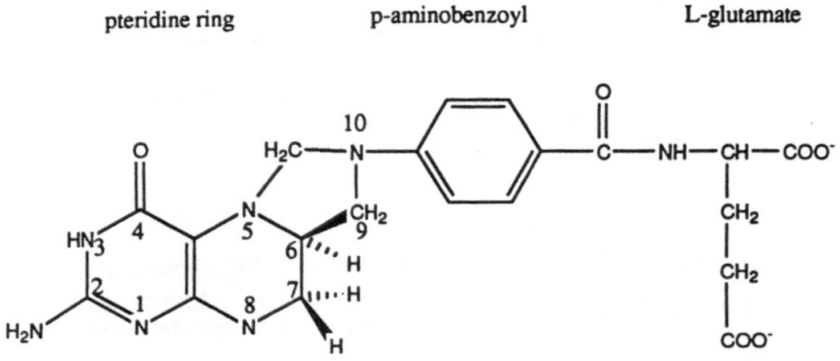

4

pteridine ring   p-aminobenzoyl   L-glutamate

5,10-methylenetetrahydrofolate

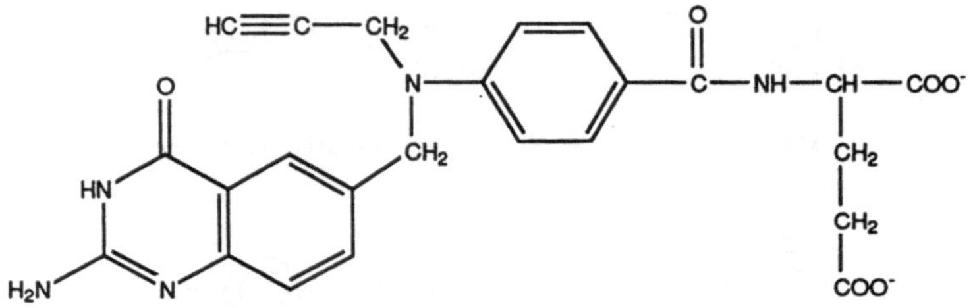

10-propargyl-5,8-dideazafolate

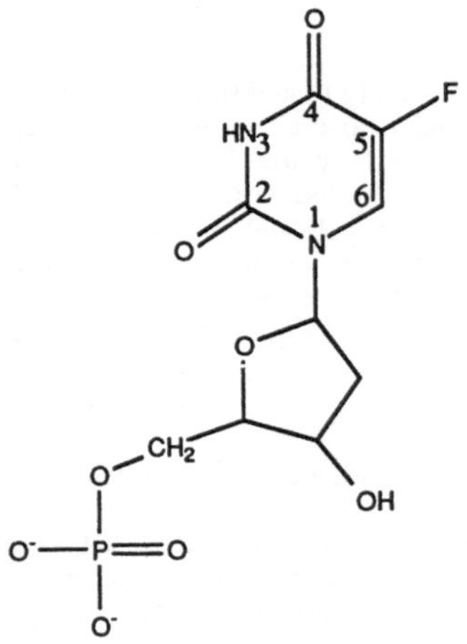

5-fluoro-2'-deoxyuridylate

Figure 3. Covalent structure and atom numbering for 5,10-methylenetetrahydrofolate (top), 10-propargyl-5,8-dideazafolate and FdUMP.

or in certain cases weaken binding to a particular class of DHFR thus providing a basis for species selectivity. Methotrexate is an example or the former. Trimethoprim, on the other hand, owes its clinical efficacy as an antibacterial to the fact that the trimethoxybenzyl substituent tightens binding to bacterial DHFR while actually reducing the inhibitor's affinity for veterbrate DHFRs compared to the parent 2,4-diaminopyrimidine.

Polar interactions between the pteridine ring of MTX and key functional groups at the active site of *L.casei* DHFR are shown schematically in Figure 5 (Bolin et al., 1982). A variety of evidence indicates that MTX when bound to DHFR is protonated (Kraut and Matthews, 1987 and references therein). The side chain of Asp 26 is nearly coplanar with the bound pteridine ring of MTX and positioned with its carboxyl oxygens approximately 2.8Å from the inhibitor's N1 and 2 amino groups respectively, suggesting the existence of two charge mediated hydrogen bonds between the enzyme and inhibitor. A second hydrogen bond exists between the 2-amino group of MTX and a fixed water molecule which is in turn hydrogen bonded to the side chain hydroxyl of Thr-116. Since this threonine is strictly conserved in all DHFR sequences, the bridging water molecule is probably an invariant feature of the enzyme structure (Bolin et al., 1982). The inhibitor's 4-amino group donates hydrogen bonds to two backbone carbonyl groups while N8 accepts a hydrogen bond from a second fixed solvent molecule.

See color insert:

Figure 4. Representation of the backbone chain folding for one subunit of *E. coli* thymidylate synthase containing bound FdUMP and 10-propargyl-5,8-dideazafolate.

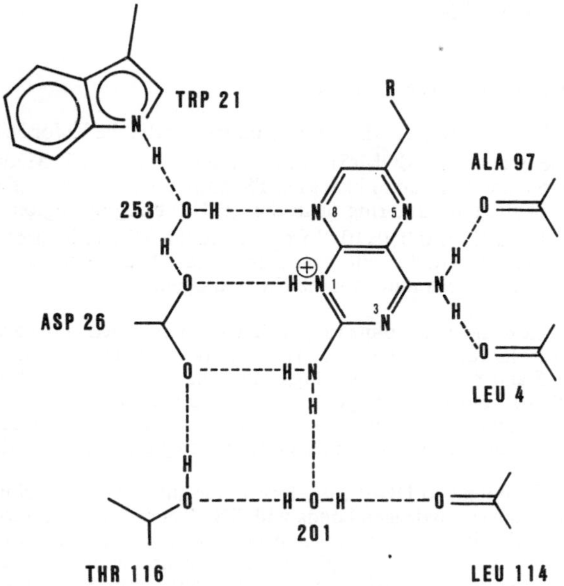

Figure 5. Schematic illustration of hydrogen bonding between *L. casei* dihydrofolate reductase and the pteridine ring of bound methotrexate. Taken from Bolin et al. (1982).

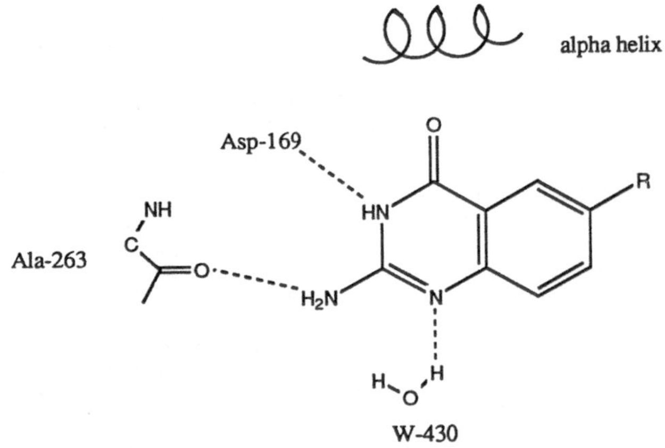

Figure 6. Schematic illustration of hydrogen bonding between *E. coli* thymidylate synthase and the quinazoline ring of bound 10-propargyl-5,8-dideazafolate.

In summary, substitution of the substrate's 4-oxo group by an amino function enhances binding to DHFR both by increasing the basicity of N1 leading to formation of charge mediated hydrogen bonds with a conserved active site carboxyl group and by providing an appropriate functional group at the 4 position which can donate hydrogen bonds to nearby protein backbone carbonyl groups.

Inhibitor Binding to Thymidylate Synthase

In contrast to DHFR, TS binds 2,4-diaminopteridines poorly, the K_i for MTX inhibition of the *E.coli* enzyme is only 19×10^{-6}M(Slavik and Zakrzewski, 1967).Modifications of the basic folic acid structure that lead to increased TS inhibition have centered primarily on the pyrazine portion of the pteridine ring or in the C9-N10 bridging region. The potent TS inhibitor, PDDF, has a K_i of 0.050×10^{-6}M against E.coli TS and incorporates changes in both regions. Compared to folate, nitrogens 5 and 8 have been replaced by carbon while a propargyl group has been substituted for the N10 hydrogen.

Figure 6 indicates how the quinazoline ring of PDDF is positioned at the active site of TS in relation to nearby polar functional groups. There are no potential hydrogen bonding groups within 5Å of the 5 and 8 ring positions suggesting that the pyrazine nitrogens of the true substrate probably do not form hydrogen bonds to nearby protein atoms. This observation may explain why quinazolines, which do not require desolvation of polar groups at these positions are better inhibitors of TS than the corresponding pteridines that do.

In contrast to MTX bound to DHFR, the four polar atoms in the pyrimidine moiety of PDDF do not make strong hydrogen bonds with TS. There is an acid group (Asp-169) in the TS active site probably weakly hydrogen bonded to N3 however the 4-oxo substituent is positioned near the edge of a buried alpha helix in a hydrophobic pocket where it cannot hydrogen bond. The pyrimidine N1 is only weakly solvated by W-430 while the 2-amino substituent makes but a single long hydrogen bond(3.1Å) to the backbone carbonyl of Ala-263.

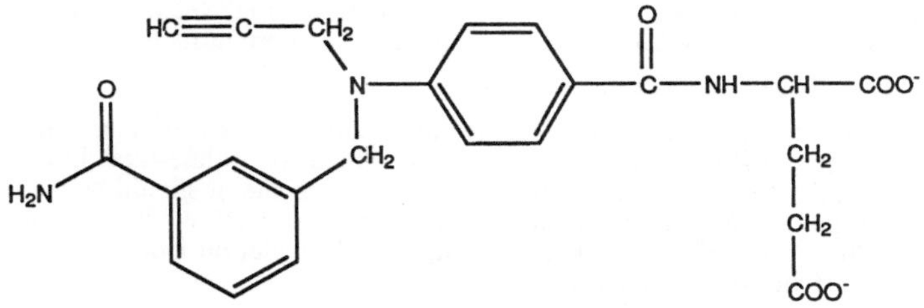

N-[p-[[(3-carboxamidophenyl)methyl]-propargylamino]benzoyl]-L-glutamic acid

Compound I

Figure 7. Nair's inhibitor of thymidylate synthase

Recently Nair and Ayling(1987) synthesized a novel disubstituted benzene(Figure 7) and reported on its properties as an inhibitor of TS. A portion of this molecule is structurally identical to part of PDDF whereas the 2-amino,4-oxopyrimidine of the latter has been replaced by a simple carboxamide. Our interest in this compound was piqued by the report that its K_i against *L.casei* TS is 1.3×10^{-6}M compared to a value of 0.086×10^{-6}M for PDDF. The corresponding values for *E.coli* TS are 0.70×10^{-6}M and 0.05×10^{-6}M respectively (private communication from Kate Welsh). What is clear from this comparison is that by removing three atoms from a portion of the quinazoline nucleus which is buried deep in the TS active site, K_i is increased only about 14 fold. Does the right hand portion of Compound I as drawn in Figure 7 bind to TS in the same manner as the corresponding portion of PDDF thus creating a gap, perhaps filled by water, at the opposite end of the active site cavity or does the smaller molecule select a completely new binding mode?

In order to address this question we crystallized *E.coli* TS in the presence of FdUMP and Compound I. Figure 8 shows a portion of the difference map between this complex and the crystallographically isomorphous TS ternary complex with FdUMP and PDDF calculated using x-ray diffraction data to a resolution of 2.4Å. Extensive negative density coincident with the position for N1, C2, and the 2-amino group of bound PDDF plus the absence of other significant features in the difference map clearly demonstrates that both compound I and PDDF bind in a similar manner to the active site of TS. A small weak negative feature adjacent to the 4-oxo position suggests that this atom has enhanced mobility in compound I bound to TS compared to the corresponding oxygen in PDDF. When Compound I binds to TS there is no indication that space vacated at one end of the active site owing to the missing N1, C2 and 2-amino is filled either by ordered solvent molecules or conformationally rearranged protein. The striking observation that creation of such a "gap" in a buried portion of the TS active site has only a minor effect on the tightness of inhibitor binding suggests that the pyrimidine portion of PDDF contributes very little net free energy of

---

See color insert:

Figure 8. Difference map between *E. coli* thymidylate synthase ternary complexes containing FdUMP and respectively 10-propargyl-5,8-dideazafolate and Compound I. Broken contours in red are at -4 sigma.

binding. This result is all the more remarkable when contrasted to the case of DHFR where an appropriately substituted pyrimidine is the single most important factor in eventuating tight binding.

These conclusions present an intriguing challenge to scientists interested in structure based design of TS inhibitors. By re-engineering the left hand part of "folate-like" inhibitors to better complement the molecular architecture of the TS active site, it should be possible to realize even more potent TS inhibition than that observed with PDDF while simultaneously minimizing binding to DHFR, thereby decoupling dTMP depletion from a block on all tetrahydrofolate-mediated one carbon metabolism.

ACKNOWLEDGEMENTS

We would like to thank Dr. Kate Welsh for supplying purified *E. coli* TS, Dr. Stuart Oatley (deceased) for his work on crystallographic refinement of the TS:FdUMP:PDDF ternary complex and Dr. M. G. Nair for his kind gift of Compound I.

REFERENCES

Bolin, J.T., Filman, D.J., Matthews, D.A., Hamlin, R.C., and Kraut, J. (1982). J. Biol. Chem., 257, 13650.

Franklin, A.L., Stokstad, E.L.R., Belt, M., and Jukes, J.H. (1947). J. Biol. Chem. 169, 427.

Hardy, L.W., Finer-Moore, J.S., Montfort, W.R., Jones, M.O., Santi, D.V., and Stroud, R.M. (1987). Science 235, 448.

Hitchings, G.H., Elion, G.B., VanderWerff, H., and Falco, E.A. (1948). J. Biol. Chem. 174, 765.

Jones, T.R., Calvert, A.H., Jackman, A.L., Eakin, M.A., Smithers, M.J., Betteridge, R.F., Nwell, D.R., Hayter, A.J., Stocker, A., Harland, S.J., Davies, L.C., and Harrap, K.R. (1985). J. Med. Chem. 28, 1468.

Kraut, J. and Matthews, D.A. in Biological Macromolecules and Assemblies Vol III pp1-71 (F. Jurnak and A. McPherson, eds.) John Wiley & Sons, New York (1987).

Matthews, D.A., Appelt, K., and Oatley, S.J. (1989a). Advances in Enz. Reg. in press.

Matthews, D.A., Appelt, K., and Oatley, S.J. (1989b). J. Mol. Biol. 205, 449.

Nair, M.G. and Ayling, J.E. (1987). Fed Proc. 46, 2036.

Osborn, M.J., Freeman, M., and Huennekens, F.M. (1958). Proc. Soc. Exp. Biol. Med. 97, 429.

Seeger, D.R., Smith, J.M., Jr., and Hultquest, M.E. (1947). J. Am. Chem. Soc. 69, 2567.

Slavik K. and S.F. Zakrzewski (1967). Mol. Pharmmacol. 3, 370.

Crystallographic and Pharmacological Studies of Antiviral Agents Against Human Rhinovirus

THOMAS J. SMITH, JOHN BADGER, MARCIA KREMER,
MARCOS OLIVEIRA, MICHAEL G. ROSSMANN, MARK A.
MCKINLAY, GUY D. DIANA, DANIEL C. PAVEAR, FRANK
J. DUTKO, ROLAND R. RUECKERT, BEVERLY HEINZ,
and DEBORAH SHEPARD

Introduction

Picornaviruses are small RNA containing viruses that cause diseases in mammals such as polio, foot-and-mouth disease, and the common cold. For reasons not yet fully understood, some members of this family have a small number of serotypes while others have many (e.g. poliovirus has three serotypes while rhinovirus has 100 serotypes). One serotype is immunologically distinct from another such that an animal can become immune to one serotype but is still susceptible to infection by another. For this reason, vaccines have been developed to prevent poliovirus infection, but the great number of rhinovirus serotypes has thwarted the development of a rhinovirus vaccine. Therefore, in the case of rhinovirus, the only hope for a "cure" seems to lie in a pharmaceutical approach.

The first stage of a picornavirus infection is recognition of specific cell surface proteins on mammalian cells. In the case of human rhinovirus, there seems to be two distinct proteins involved (Abraham and Colonno, 1984). The different rhinovirus serotypes have been grouped according to which of the two proteins they recognize. These two groups are called the major and minor groups (the names were derived from the number of serotypes falling into each of the two groups). Recent experiments have identified the major group cell surface protein as a immunologically important cell adhesion protein called ICAM-1 (Greve et al, 1989; Stauton et al, 1989).

9

10

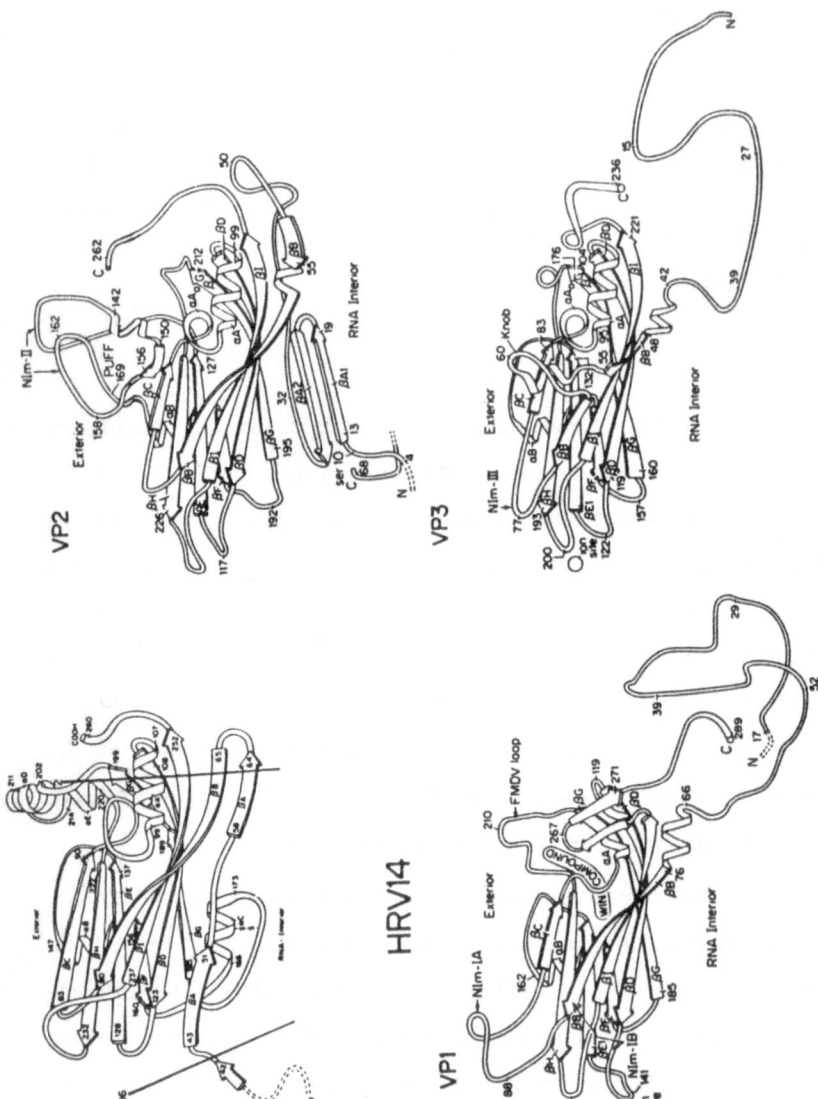

Figure 1. Ribbon drawings of the eight stranded, anti-parallel, ß-barrel structures of VP₁, VP₂, and VP₃ from human rhinovirus 14 compared to the coat protein from southern bean mosaic virus (SBMV).

The next stage of a rhinovirus infection is invagination of the virus particle via a coated pit mechanism and transport into the cell. Upon entry into the cytoplasm, the interior of the receptosomal vesicle undergoes acidification. It has been postulated that this drop in pH induces a conformational change in the capsid which is required to release the RNA into the cytoplasm. Similar pH effects have been seen with influenza virus (for a review, see White *et al*, 1983) Once the RNA is inside the cytoplasm, the viral genetic material takes over the host protein synthesis machinery and viral replication starts.

The structure of human rhinovirus 14 has been determined to atomic resolution (Rossmann *et al*, 1985). The 300Å icosahedral shell is composed of four viral proteins: VP_1, VP_2, VP_3, and VP_4. The first three proteins form eight stranded, anti-parallel, ß-barrels whereas the small VP_4 protein lies at the interface between the RNA interior and the protein capsid (fig. 1). The basic structural motif of an eight stranded, anti-parallel, ß-barrel is conserved not only amongst these three rhinovirus proteins but was also found in spherical plant viruses (tomato bushy stunt virus [Harrison *et al*, 1978], southern bean mosaic virus [Abad-Zapatero *et al*, 1980], cowpea mosaic virus [Stauffacher *et al*, 1987],and bean pod mottle virus [Chen *et al*, 1989]), an insect virus (black beetle virus [Hosur *et al*, 1985]), and other members of the picornavirus family (poliovirus [Hogle *et al*, 1985], foot-and-mouth disease virus [Acharya *et al*, 1989], mengovirus [Luo *et al*, 1987], and human rhinovirus 1A [Kim *et al*, submitted]). The major differences between these spherical RNA viruses lie on the surfaces or "decorations" of the ß-barrels. In the case of human rhinovirus 14, there is a 25Å deep "canyon" encircling each icosahedral five-fold axis. At the solvent exposed rim of the canyon, lie the immunogenic sites of the virus. Because of the dimensions of the canyon, and the position of the immunogenic sites, it was postulated that the bottom of the canyon may be used to recognize the cell surface receptor proteins (fig. 2). This would allow the virus to prevent

12

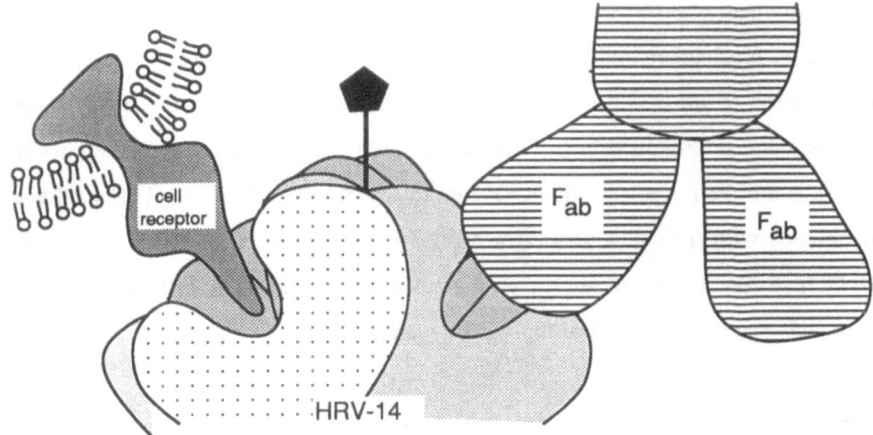

Figure 2. Schematic drawing representing the canyon hypothesis.

antibody binding by altering the decorations of ß-barrels without having to change the portion of the capsid necessary for cell recognition.

Results and Discussion

The Sterling Research Group has synthesized a series of compounds that inhibit picornavirus infection. The basic structure of compounds (WIN compounds) is an isoxazole ring connected to oxazolinylphenoxy system via an aliphatic chain. The structure of the compounds studied crystallographically thus far are shown in fig. 3. These agents are not virucidal, in that the compounds bind reversibly and do not permanently inactivate the virus. Early experiments demonstrated that these hydrophobic compounds do not affect the ability of poliovirus or human rhinovirus-2 to bind to the cell surface proteins but rather affect the ability of the virus to uncoat and release RNA into the cytoplasm (Fox *et al*, 1986). This family of hydrophobic compounds was thought to cause this effect by binding to the capsid causing a stabilization effect, thereby preventing the conformational changes. This mechanism was supported by studies that demonstrated that these WIN compounds partially protect the capsids against thermal and acid inactivation (Diana *et al*, 1985). These compounds have a broad specificity; they inhibit polioviruses, rhinoviruses, echo-9 virus, and coxsackievirus. The most exciting results with human poliovirus demonstrated that these compounds are also effective *in-vivo* (McKinlay and Steinberg, 1985). Cerebral injections of human poliovirus were administered to mice. After the inoculations, oral administration of the WIN compounds effectively prevented the onset of paralysis. This demonstrated that the compounds are effective *in-vivo*.

When these crystallographic studies were initiated, neither the site nor stoichiometry of binding was known. Since WIN compounds are hydrophobic, there was some concern as to whether they would be able to intercalate into crystals of HRV-14 so that difference fourier techniques could be applied. Crystals of HRV-14 were soaked with radioactive WIN compound to determine whether or not the compounds could penetrate the crystal matrix and what the stoichiometry of binding was. It was found that approximately 60 molecules of WIN 51,711 (compound IV, fig. 2) bound per virion without disrupting the crystals (Smith *et al*, 1986). These results were later confirmed with non-crystalline HRV-14 using the spin sephadex method of Penefsky (1977). These results demonstrated that, in solution or in the crystal, approximately one molecule of drug bound per protomeric unit (VP_1, VP_2, VP_3, and VP_4 together).

Since it was clear that the drug was able to intercalate into the crystals, unlabeled WIN 51,711 and WIN 52,084 (compound I(S) and I(R) fig. 2) were soaked into crystals of HRV-14 and data were collected at the Cornell High Energy Syncro-

14

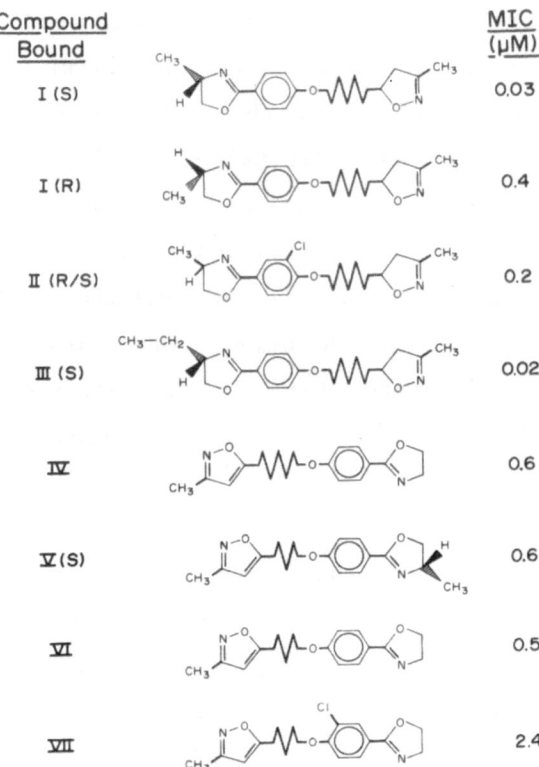

Compound Bound		MIC (μM)
I (S)		0.03
I (R)		0.4
II (R/S)		0.2
III (S)		0.02
IV		0.6
V (S)		0.6
VI		0.5
VII		2.4

Figure 3. Chemical structures of some of the WIN compounds studied crystallographically to date. The orientation of the compounds as they are bound to HRV-14 are shown here. The right hand side of the table represents the deeper recesses of the VP₁ ß-barrel and the left hand side represents the entrance to the drug binding region. Also shown are the MIC values (the concentration of compound required to inhibit plaque formation by 50% in *in-vitro* tissue culture infectivity studies).

tron Source (CHESS). Difference maps were calculated using the native phases, native structure amplitudes and new HRV14-drug complex structure amplitudes. This map was difficult to interpret because conformational changes had occurred in the capsid, resulting in many positive and negative electron density peaks. If $F_{observed}$ and α_{native} are used to calculate an electron density map, the drug and the conformational changes are not visible. This is due to the bias of the native structure in the electron density. This bias can come from two sources: a) the bias from the native phases and/or b) not all of the particles are 100% saturated with drug in the crystal. The hypothesis that the partial saturation of the virions with drug in the crystal might be a problem was subsequently borne out by a demonstrable increase in the observed saturation of the virions in the crystals when the crystal soaking times were increased from 4 days to 2 weeks. In order to better visualize the drug bound structure, a map was calculated using the following structure amplitude relationship;

$$F_{new\ structure} = \left| \ F_{observed} - (1-k)F_{native} \ \right|$$

where k equals the observed fraction of drug-virus complex in the electron density.

These resulting maps clearly showed the position of the bound drug and the conformational changes associated with drug binding (Smith *et al*, 1986). Examples of the electron density representing drug bound the the virions are shown in figures 4a,b. As shown in Figure 5a,b, these WIN compounds bind in the heart of the VP_1 ß-barrel. In order to reach the final binding site, the compounds must travel to the bottom the "canyon" then continue deeper under the floor of the canyon. As shown in figure 6a, in order to bind, these drugs must move residues 1151-1155 of the FMDV loop (so named because of its sequence homology with the antigenic loop of foot-and-mouth disease virus) out of the way. In this four number nomenclature, the first number identifies the VP protein and the residue number is given by the last three numbers. This movement in the FMDV loop causes concerted movements in other VP_1 ß-barrel strands in order to preserve the hydrogen bonding pattern (fig. 6b). Most notably, methionine 1221 blocks the entrance to the "drug binding pocket" in the native structure and is pushed out of the way when the drug binds. Histidine 1220 was moved the largest distance, and may be partially responsible for the change in pI of the capsid upon the binding of these uncharged compounds (Smith *et al*, 1986). Even though there are large differences in the substituents on the rings of the compounds examined thus far (fig 2) this conformational change is the same (within experimental error) no matter which one is bound.

The environment of the drug binding region is very hydrophobic (fig. 7a,b). The only polar interaction between the capsid protein and the WIN compounds is between asparagine 1219 and the nitrogen of either the isoxazole or oxazoline rings (depending upon the particular compound). This interaction may not be imperative since some of the compounds subsequently studied have a rather long hydrogen bonding distance to asparagine 1219. Even though the basic structure for the compounds examined is the same, they bind in different orientations. From this list, tentative rules for orientation can be derived. If the aliphatic chain is five carbons long, then the drug binds with the oxazolinylphenoxy ring system deepest in the pocket, no matter what substituents are on the ring structures. All of these shorter compounds favor the deeper, more hydrophobic portion of the ß-barrel. If, on the other hand, the aliphatic chain is seven carbons long, substituents on the oxazoline ring do determine the orientation of the compound. If there are substituents on the oxazoline ring (e.g. methyl, ethyl, and propyl), then the compound will enter wiht the isoxazole group towards the deepest end of the pocket. Without a substituent on the oxazoline ring, the compound might enter either way but probably prefers the oxazoline-phenoxy end first. Curiously, different WIN compounds can bind in opposite orientations (suggesting a great deal of freedom in the binding modes), yet sometimes subtle differences in the compounds can greatly influence activity. When the aliphatic chain is seven carbons long, and the oxazoline ring is methylated, the S isomer is over 10 times more active than the R isomer. This is probably due to favorable interactions, in the case of the S isomer, with leucine 1106 and serine 1107 (Diana *et al*, 1988).

A

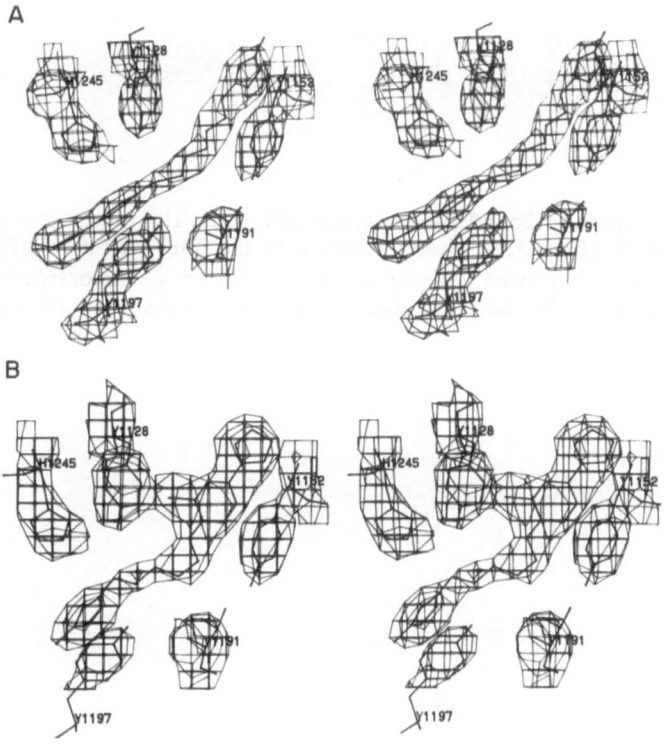

B

Figure 4. Electron density representing WIN compounds I(S) and VII (from figure 2) as bound to HRV-14.

The effect of the WIN compounds on the stability of the capsid can be explained by entropy effects.   When hydrophobic substances are in an aqueous environment, water molecules are forced to be ordered about it. This leads to an unfavorable entropy term. When the drugs move from the aqueous solution to the hydrophobic environment of the drug binding pocket, the ordered waters are shed, making the virion a more disordered state, or a more favorable entropy term. In other words, denaturation of a drug-virion complex is more unfavorable

Figure 5. Position of bound drug in protomeric unit. This figure shows a C–α stereo drawing of the four viral proteins with the drug bound to VP$_1$. The protomeric unit is viewed with the exterior at the top of the drawing and the RNA interior at the bottom. The icosahedral five-fold axis is on the right hand side.

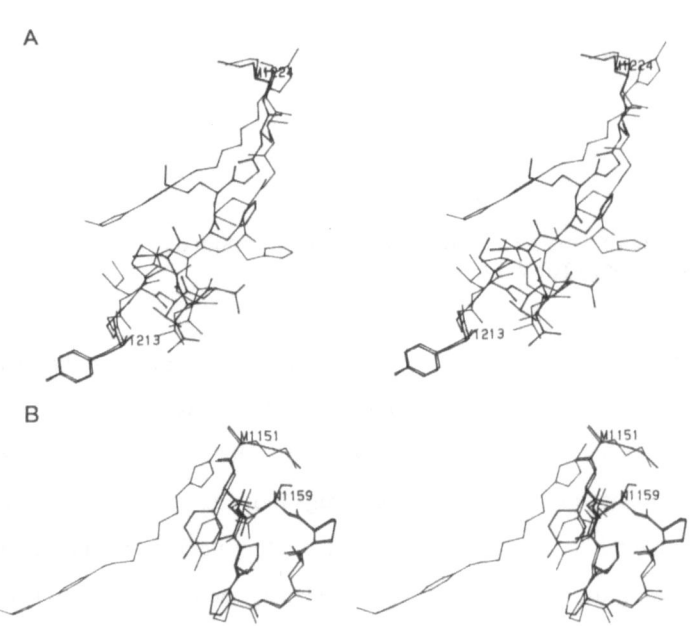

Figure 6. Conformational changes induced by WIN compound binding. The protein is labelled according to the four number system. The native conformation is shown in heavier lines and the drug induced conformation is shown in the lighter lines.

A

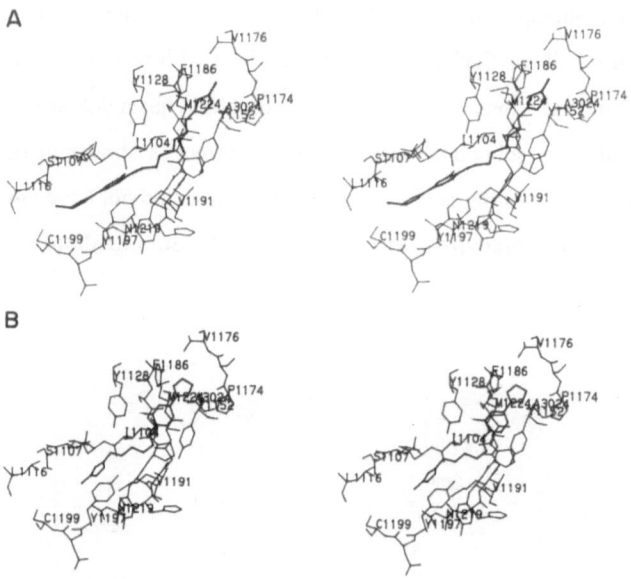

B

Figure 7. Environment of the drug binding site. The drug is shown in heavy lines and the protein is shown in thinner lines. Figure 7a shows the conformation of compound II (R/S) and the surrounding protein, and figure 7b shows the conformation of the smaller (n=5) compound VII.

than a virion alone because not only must the hydrophobic drug pocket become solvated, but the hydrophobic WIN compound must also be extruded into the environment. A similar effect was was seen when T4 lysozyme was stabilized by filling, internal, hydrophobic cavities with large hydrophobic sidechains (Matthews *et al*, 1987).

Recent studies with HRV-14 have shown that, in addition to the drug's stabilizing effect on the capsid, the drugs also affect the ability of the virus to bind to cell surface receptors (Pavier *et al*, 1989). As the concentration of the drug increases, the number of virions that can bind to cell membrane preparations drop from 100% to 5-20% with the more efficacious drugs. Since we know that

20

the conformational changes upon drug binding are limited to the very bottom of the canyon, then it follows that the canyon is involved with receptor recognition and that it is the deformations caused by drug binding that has a deleterious effect on receptor binding. A schematic drawing of this effect is shown in figure 8. What is particularly interesting is that the WIN compounds do not have any effect on receptor binding on rhinovirus serotypes 1A and 2. Both of these serotypes

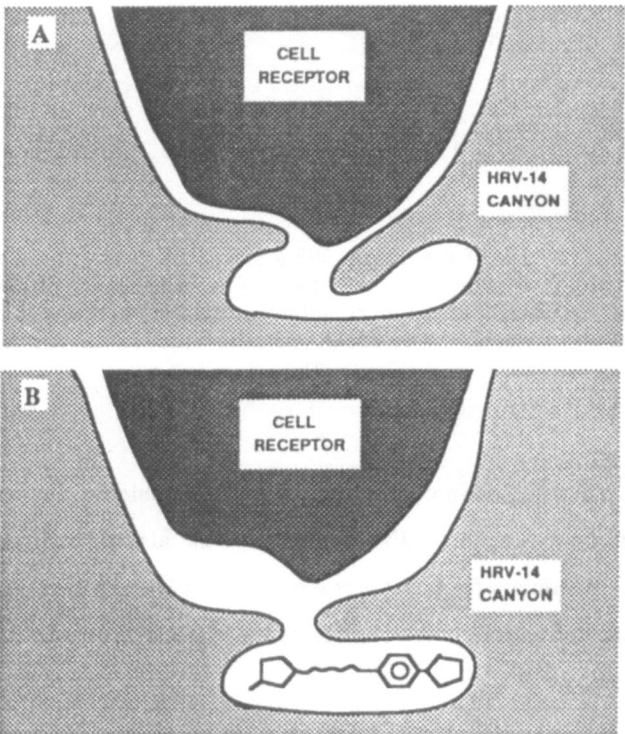

Figure 8. Schematic drawing of the WIN compound's effect on HRV-14 binding to cell membranes. Figure 8a shows the normal interaction between the receptor protein and HRV-14 whereas figure 8b shows how the drug induced changes in the canyon floor might interfere with receptor binding. In figure 8b, the movement in the canyon floor prevents the receptor from binding to the bottom of the canyon and making the same kind of complemetary fit as is seen in figure 8a.

belong to the minor group of rhinoviruses rather than the major group of receptors of rhinovirus. One interpretation of these results is that the minor group cell receptor protein does not interact with the FMDV loop (which moves upon drug binding) in the same way as the major group does, and therefore the minor group of rhinoviruses are insensitive to the drug induced conformational changes in the canyon floor. Some support for this hypothesis comes from crystallographic results which show that the conformation of the native HRV-1A canyon floor might be more like the conformation of the drug-bound HRV-14 structure (Kim *et al*, submitted).

Drug Resistant Mutants

RNA viruses are infamous for their lack of replication fidelity. When HRV-14 replicates, the RNA of many of the progeny are different, or have errors. If neutralizing antibodies or antiviral agents are added to the infected cells, then those errors lending the virions resistance to the agents are selected for. In the case of the WIN compounds two general mutations can be selected for: virions that can resist high concentrations of drug and mutants that can only resist low concentrations of drug. Fifty-six mutants that are resistant to high concentrations of WIN 52,084 (compound I, figure 2) have been selected for and sequenced (Heinz *et al*, in press, 1989). Out of all of the possible residues that can be mutated in the drug binding pocket, only two residues were found to mutate naturally in the highly resistant mutants: valine 1188 and cysteine 1199. These residues mutated to several types of new side chains - all larger. Examples of both mutations were crystallized and their structures determined. No other change in protein structure was observed. On the other hand, the cysteine to tyrosine mutation at residue 1199 did cause some conformational rearrangement. This is due to the lack of space about the cysteine to accommodate the larger side chain. Preliminary binding studies demonstrated that neither mutation was able

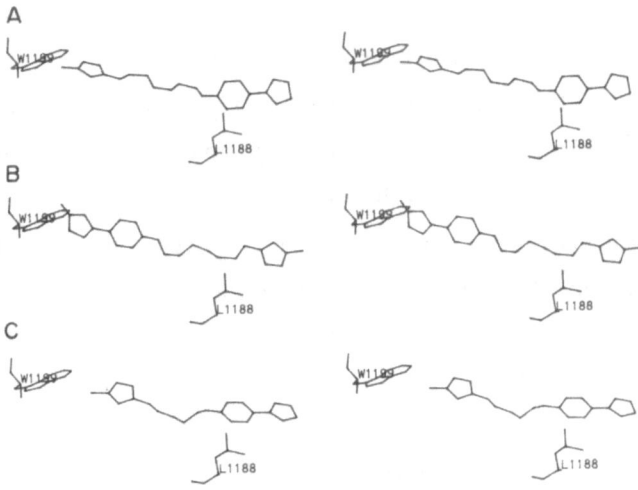

Figure 9. Stereo drawings of two high resistant mutants (C1199T and V1188L) with various WIN compounds modelled in their bound conformations. Figure 9a shows the close contacts that interfere with drug binding in the case of compound IV, and 9b shows the same for compound I(S). Figure 9c demonstrates that there might be enough room for a shorter, n=5, compound (compound VI).

to bind drug well, and from examination of the structures clearly showed that the drugs do not have enough space to bind. Figure 9 shows the results of modelling one of the other mutations (C1199T) based on the C1199Y structure. One interesting result from examining these modelling studies was that while the longer compounds (with an aliphatic chain of n=7) cannot bind in the cavity without colliding with the tryptophan at 1199, there may be enough room to accommodate a shorter (n=5) compound (fig. 9c). Indeed, the shorter compound was more effective against this particular mutant (Badger *et al*, 1988). If WIN compounds become used clinically, these studies have shown what the most probable naturally occurring mutants will be (at least in the case of HRV14) and how the pharmacologists might deal with them.

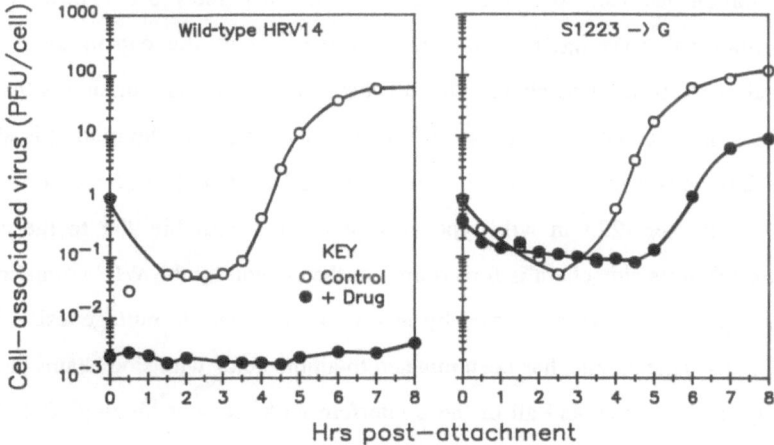

Figure 10. Single cycle growth curves of virus in HeLa cell suspensions in the presence (filled circles) and the absence (open circles) of WIN 52,084 (compound I(S), figure 3). (Left) Wild type virus. (Right) The low-resistant mutant S1223G. Virus was complexed with drug (2μg/ml if WIN 52,084 in 0.1% dimethyl sulfoxide and 0.4% bovine serum albumin) by incubating for 1 hour at 20°C, then overnight at 4°C. After an attachment (10 pfu/cell) period of 30 min at room temperature in medium containing drug, cells were sedimented to remove free virus, resuspended in fresh medium containing drug, and incubated at 35°C. Aliquots were removed at intervals and stored frozen at -70°C; cell-associated virus as released by 3 cycles of freezing and thawing, and infectivity was titered by plaque assay. Infectivity recovered at time zero is a measure of the amount of virus attached to cells. It is evident from the curve that drug inhibits the attachment of wild-type virus 1000-fold; in contrast, it inhibits the attachment of the Gly-1223 mutant only 4-fold. The observation that both wild-type and mutant viruses are strongly thermostabilized by drug (Heinz, Shepard and Rueckert, in press) demonstrates that the drug is complexed with both viruses when present at the concentration used in this experiment (data not shown).

The second class of mutants is comprised of low-resistance mutants (Heinz *et al.*, 1989, in press). One mutant that has been identified is Ser-1223 -> gly. This mutation is found on the FMDV loop and lies on the canyon floor above the drug-binding pocket. This region moves upon drug binding. The structure of this mutation has been solved, and the difference map showed that little or no conformational rearrangement occurred in the rest of the capsid upon the removal of the serine side chain. One notable property of this mutation is that it alleviates the attachment-inhibitory effect of bound drug; i.e. the mutant is able to attach to cells even when drug is bound (Figure 10). One interpretation of this result is that ser-1223 in wild-type virus interferes with binding to the cell receptor when its side chain is forced up into the canyon by the WIN compound. This suggestion is also supported by results of site-specific mutagenesis. By chance, this same residue has been mutated to amino acids with side chains larger than gly (ala,thr, asn), and all of these interfere with receptor binding (Colonno et al.,1988). However, the observation that WIN 52084 prolongs the eclipse period of the ser-1223 -> gly mutant, even though the attachment-inhibitory effect of the drug has been largely attentuated, implies that the drug has a second inhibitory effect in HRV14. While there is evidence that WIN compounds can bind to this particular mutant, studies are underway to make sure that the Kd's for the WIN compounds are comparable to wild type. This will serve as a control to make sure that the above effects are not merely due to binding affinity differences.

Conclusions

The antiviral WIN compounds bind to the internal portion of the VP1 ß-barrel and inhibit infectivity via two mechanisms; inhibition of infectivity by stabilization of the capsid by hydrophobic effects and inhibition of cell attachment by altering the conformation of the canyon floor. In the case of human rhinoviruses 1A and 2 (both members of the minor group), only the stabilization of the capsid is seen. However, HRV-14 (a major group rhinovirus) is sensitive to both modes of inhibition. Studies are in progress to determine whether this attachment effect occurs when WIN compounds are bound to all major group rhinoviruses, but not when WIN compounds are bound to minor group rhinoviruses. Since we know that the conformational changes induced by drug binding are limited to the bottom of the canyon floor, these results are consistent with the canyon hypothesis. If these drug induced effects on cell membrane binding are seen amongst all of the major group rhinoviruses, then we may also deduce that at least one difference between the major and minor groups are in the interactions between their cell surface proteins and the residues around 1223 of the FMDV loop. The role of this specific serine residue in cell membrane binding has been shown by site specific mutation studies (Colonno, 1988) as well as a low-resistant, naturally occurring mutant grown in the presence of WIN 52,084.

These studies have given some hope as well as realistic pessimism to medicinal chemists trying to design compounds that will bind to a target protein of known structure. All of the various members of this family of compounds induce nearly identical conformational change upon binding. This will help future computational experiments since the conformation of the drug bound conformation can be assumed. On the other hand, it would have been nearly impossible (because of the closed drug pocket in the native structure) to have guessed that a compound would be able to intercalate into the heart of the VP1 ß-barrel by merely looking at the native structure. Therefore, *de-novo* drug design based on a native structure alone, will probably always be difficult because of unexpected conformational changes. These studies also lend a cautionary note to drug design since it was not intuitively obvious that a compound binding to the inner portions of the VP_1 protein would be sufficient to cause enough stabilization to inhibit infectivity.

Acknowledgements

This work was supported by a Jane Coffin Childs grant to TJS. MGR and RRR were supported by grants by NSF, NIH and the Sterling Research Group . We are also grateful for the kind assistance given by the personnel at the Cornell, Stanford, Brookhaven, and Daresbury syncrotrons.

References

Abad-Zapatero, C., S. S. Abdel-Meguid, J. E. Johnson, A. G. W. Leslie, I. Rayment, M. G. Rossmann, D. Suck, and T. Tsukihara., Structure of southern bean mosaic virus at 2.8Å resolution. Nature 286:33-39.

Abraham, G. and R. J. Colonno, Many rhinovirus serotypes share the same cellular receptor. J. Virol. 51:340-345 (1984).

Acharya, R., E. Fry, D. Stuart, G. Fox, D. Rowlands, F. Brown, The Three-dimensional structure of foot-and-mouth disease virus at 2.9Å resolution. Nature 337:709-716(1989).

Badger, J., I. Minor, M. J. Kremer, M. A. Oliveira, T. J. Smith, J. P. Griffith, D. M. A. Guerin, S. Krishnaswamy, M. G. Rossmann, M. A. McKinlay, G. D. DIana, F. J. Dutko, R. R. Reuckert, and B. Heinz, Structural analysis of a series of antiviral agents complexed with human rhinovirus 14. Proc. Natl. Acad. Sci. USA 85:3304-3308 (1988).

Chen Z., C. Stauffacher, Y. Li, T. Schmidt, W. Bomu, G. Kremer, G. Lomonossoff, and J. E. Johnson, Protein nucleic acid interactions in a spherical virus: the structure of beanpod mottle virus at 3.0Å Resolution. Science (in press).

Colonno, R. J., J. H. Condra, J. Mizutani, P. L. Calahan, M. E. Davies, and M. A. Murko, Evidence for the direct involvement of the rhinovirus canyon in receptor binding. Proc. Natl. Acad. Sci. USA 85:5449-5453 (1988).

Diana, G. D., M. A. McKinlay, M. J. Otto, J. Akullian, and C. Oglesby, {[(4,5-Dihydro-2-oxazolyl)phenoxy]alkyl}isoxazoles. Inhibitors of picornavirus uncoating. J. Med. Chem. 28:1906 (1985).

Diana, G. D., M. J. Otto, M. A. McKinlay, R. C. Oglesby, A. Treasurywala, E. G. Maliski, M. G. Rossmann, and T. J. Smith, Enantiomeric effects of homologues of disoxaril on the inhibitory activity against human rhinovirus 14. J. Med. Chem. 31:540-544(1988).

Fox, M. P., M. J. Otto, and M. A. McKinlay, Prevention of rhinovirus and poliovirus uncoating by WIN 51,711, a new antiviral drug. Antimicrob. Agents Chemother. 30:110 (1986).

Greve, J. M., G. Davis, A. M. Meyer, C. P. Forte, S. Connolly Yost, C. W. Marlor, M. E. Kamarck, and A. McClelland, The Major Human Rhinovirus Reptor is ICAM-1. Cell 56:839-847.

Harrison, S. C., A. J. Olson, C. E. Schutt, F. K. Winkler, and G. Bricogne, Tomato bushy stunt virus at 2.9Å resolution. Nature 276:368-373 (1978).

Heinz, B. A., R. R. Reuckert, D. A. Shepard, F. J. Dutko, M. A. McKinlay, M. Fancher, M.G. Rossmann, J. Badger, T. J. Smith, Genetic and Molecular analysis of spontaneous mutants of human rhinovirus 14 resistant to an antiviral compound (J. Virol. in press).

Heinz, B. A., D. A. Shepard and R. R. Rueckert. Escape mutant analysis of a drug-binding site can be used to map functions in the rhinovirus capsid. In: W. G. Laver and G. Air (eds.), The use of X-ray crystallograpohy in design of antiviral agents. Academic Press, Inc., San Diego, in press.

Hosur, M. V., T. Schmidt, R. C. Tucker, J. E. Johnson, T. M. Gallagher, B. H. Selling, and R. R. Reuckert, The structure of an insect virus at 3.0Å resolution. Proteins Struct. Function Genet. 2:167-176 (1987).

Kim, S., T. J. Smith, M. S. Chapman, M. G. Rossmann, The crystal structure of human rhinovirus serotype 1A (HRV1A). Journ. Mol. Biol. (submitted).

Luo, M., G. Vriend, G. Kamer, I. Minor, E. Arnold, M. G. Rossmann, U. Boege, D. G. Scraba, G. M. Duke, A. C. Palmenberg, The Atomic Structure of Mengo Virus at 3.0Å resolution. Science 235:182-191 (1987).

Matthews, B. W., H. Nicholson, and W. J. Becktel, Enhanced protein thermostability from site directed mutations that decrease entropy of unfolding. Proc. Natl. Acad. Sci. USA 84:6663-6667 (1987).

McKinlay, M. A., and B. A Steinberg, Oral efficacy of WIN 51,711 in mice infected with human poliovirus. Antimicrob. Agents Chemother. 29:30 (1985).

Penefsky, H. S., Reversible binding of Pi by beef heart mitochondrial adenosine triphosphotase. Journ. Biol. Chem. 252:2891-2899 (1977).

Pevear, D. C., M. J. Fancher, P. J. Felock, M. G. Rossmann, M. S. Miller, G. Diana, A. M. Treasurywala, M. A. McKinlay, and F. J. Dutko, Conformational change in the floor of the human rhinovirus canyon blocks adsorption to Hela cell receptors. Journ. Virol. 63: 2002-2007 (1989).

Rossmann, M. G., E. Arnold, J. W. Erickson, E. A. Frankenberger, J. P. Griffith, H. J. Hecht, J. R. Johnson, G. Kamer, M. Luo, A. G. Mosser, R. R. Rueckert, B. Sherry, and G. Vriend, Structure of a human common cold virus and functional relationship to other picornaviruses. Nature 317:145-153 (1985).

Smith, T. J., M. J. Kremer, M. Luo, G. Vriend, E. Arnold, G. Kamer, M. G. Rossmann, M. A. McKinlay, G. D. Diana, and M. J. Otto, The site of attachment in human rhinovirus 14 for antiviral agents that inhibit uncoating. Science 233:1286-1293 (1986).

Stauffacher, C. V., R. Usha, M. Harrington, T. Schmidt, M. V. Hosur, and M. V. Johnson, The structure of cowpea mosaic virus at 3.5 Å resolution. In Crystallography in Molecular Biology, D. Moras, J. Drenth, B. Strandberg, D. Suck, and K. Wilson, eds., pp. 293-308 (1987).

Staunton, D. E., V. J. Merluzzi, R. Rothlein, R. Barton, S. D. Marlin, T. A. Springer, A cell adhesion molecule, ICAM-1, is the major surface receptor for rhinovirus. Cell 56:849-853 (1989).

White, J., M. Keilian, and A. Helenius, Membrane fusion proteins of enveloped animal viruses. Q. Rev. Biophys. 16:151-195 (1983).

Structural Studies of Elastase-Inhibitor Complexes with ß-Lactams

MANUEL A. NAVIA, BRIAN M. MCKEEVER,
and JAMES P. SPRINGER

INTRODUCTION

Emphysema is a chronic disease of the lower respiratory tract, characterized by the destruction of the alveolar walls of the lung (Robbins, *et. al.*, 1984). It is a progressively degenerative disease for which there is no effective therapy at present. Human neutrophil elastase (HNE) is a serine protease which is stored in the azurophilic granules of neutrophils, and which has broad specificity for small aliphatic groups at the P_1 position of substrates (Powers *et. al.*, 1986). HNE digestion of elastin, the principal structural protein in the alveoli, is believed to be the molecular event that underlies the clinical manifestation of emphysema (Janoff, 1985). HNE is a single chain glycoprotein of 27,000 daltons molecular weight (Sinha *et. al.*, 1987) which is synthesized during the development of the mature neutrophil (Clark *et. al.*, 1984). Normally, proteolytic activity is released intracellularly within the neutrophil after the azurophilic granules entraining the enzyme are discharged into phagosomes containing engulfed debris (Parmley *et. al.*, 1986). Tissue damage results when enzyme inevitably leaks out of the neutrophil (and into the extracellular space) by regurgitation during phagocytosis or by cell death, for example. To counter this dangerous condition, proteinaceous inhibitors have evolved that can rapidly inactivate HNE and arrest the process of tissue destruction. Some idea of the importance of these inhibitors is given by their concentration in serum, which is normally around 1.3 grams per liter for the $alpha_1$ proteinase inhibitor (Carrell *et. al.*,1982). Health problems are noted when these concentrations drop below about 35% of normal (Carrell, 1986). Because $alpha_1$ proteinase inhibitor is sensitive to oxidation, it has been suggested that the correlation between smoking and emphysema might be the result of an imbalance between free HNE and its inhibitors in the lung (Carp *et. al.*, 1982).

One therapeutic approach to the problem of HNE-mediated damage has been direct replacement, by inhalation into the lung, of recombinant $alpha_1$ proteinase inhibitor (Gadek *et. al.*, 1983). As an alternative strategy, we have been involved in the development of small molecular weight inhibitors of HNE which would be more like conventional pharmaceutical agents. Many tissue component proteins other than elastin have been shown to be substrates for HNE, including proteoglycans and various collagen types. As such, many other degenerative diseases, including rheumatoid arthritis, might involve free HNE activity in some way (Janoff, 1978). Definitive data on the role of HNE in other diseases has been difficult to obtain, given the lack of even tolerable HNE inhibitors for use in chronic animal models (Ranga *et. al.*, 1981). HNE may also be involved in acute emergency medical conditions such as the adult respiratory distress syndrome (Cochrane *et. al.*, 1983).

ELASTASE INHIBITION BY ß-LACTAMS

One novel class of elastase inhibitors under investigation at Merck are the ß-lactams (Doherty *et. al.*, 1986). These compounds are best known for their therapeutic role as antibiotics, where they are believed to target the essential enzymes of bacterial cell wall synthesis, the so-called penicillin binding proteins (Waxman *et. al.*, 1983). Bacteria can acquire resistance for their own defense by producing an assortment of enzymes that can hydrolyze ß-lactam antibiotics. These soluble ß-lactamases have been extensively studied biochemically, both to gain a theoretical understanding of ß-lactam - enzyme interactions (see *eg.* Knowles, 1985), and to more directly attack the serious clinical problem of antibiotic resistance (Bush, 1988). As with HNE, the

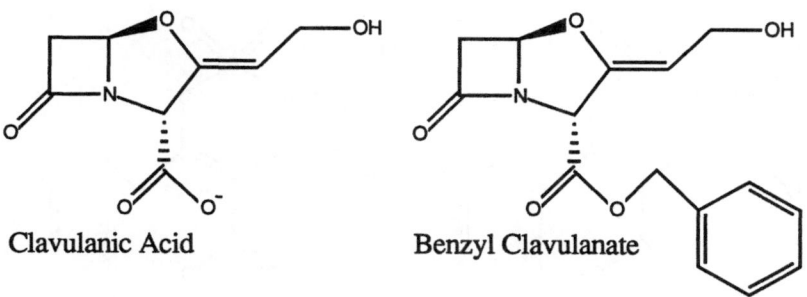

Clavulanic Acid   Benzyl Clavulanate

Figure 1. Structures of clavulanic acid and benzyl clavulanate. The former was shown to be a potent ß-lactamase inhibitor (reviewed in Cole, 1981). The latter was the first of the ß-lactam elastase inhibitors to be discovered at Merck (unpublished work of Dr. M. Zimmerman).

mechanism of action of a number of these ß-lactamases is proposed to include acylation of substrate by an active site serine. Recent structural information (Herzberg *et. al.*, 1987) shows, however, that this residue does not form part of a catalytic triad as found in the trypsin-like family of serine proteases of which HNE and porcine pancreatic elastase (PPE) are members. Given that the ß-lactam antibiotics inhibit the penicillin binding proteins by a mimic of their D-amino acid substrates, it is not surprising that cephalothin (a conventional ß-lactam antibiotic) does not inhibit the elastases, which prefer L-amino acids as their substrates (Doherty *et. al.*, 1986).

Historically, the initial discovery of ß-lactam inhibition of elastase emerged indirectly from the elegant research on ß-lactamase inhibition that took place in the late 70's. This work led to the formulation of various mechanisms by which inhibition of those enzymes might take place (see *eg.* Knowles, 1985 for a comprehensive review). Two benzyl esters of the ß-lactamase inhibitors clavulanic acid and sulbactam (penicillanic acid sulfone) proved to be time-dependent inhibitors of elastase, although the free acids (required for ß-lactamase activity) were inactive. Actually, a double serendipity was involved, since in the penicillin 6-position (corresponding to the cephalosporin 7-position), the elastases prefer alpha-substituents (and very

32

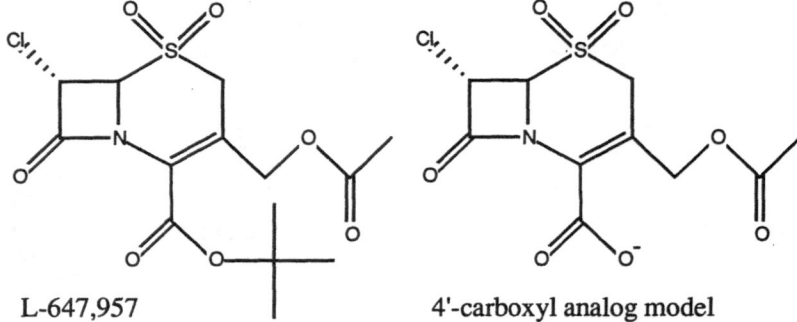

L-647,957   4'-carboxyl analog model

Figure 2. Structure of L-647,957 (3-acetoxymethyl-7-alpha-chloro-3-cephem-4-carboxylate-1,1-dioxide tert-butyl ester), an irreversible inhibitor of human neutrophil and porcine pancreatic elastases. The more antibiotic-like 4'-carboxyl analogs are ineffective as ß-lactam elastase inhibitors, as explained in the text and figure 5.

poorly tolerates beta-), whereas the ß-lactamases prefer beta. As it turns out, both benzyl clavulanate and the sulbactam benzyl ester, are unsubstituted in the 6-position and are therefore acceptable to both enzymes (see figure 1). Timely development of these leads by Doherty et. al. (1986) led to a cephalosporin sulfone series of potent inhibitors of both human neutrophil elastase, the actual drug target of interest, and porcine pancreatic elastase, a convenient model enzyme for HNE.

L-647,957 (figure 2) is the most potent of the ß-lactam elastase inhibitors reported by Doherty et. al. (1986). We have reported the three-dimensional structure of a complex of PPE inhibited by L-647,957 (figure 3). PPE was used in these studies instead of HNE because it is easier to obtain, crystallizes readily (leaving the active site of the enzyme exposed to solvent), and diffracts to high resolution (Sawyer et. al., 1978). Compounds soaked into native PPE crystals react directly with the crystalline enzyme, and allow one to examine the resulting enzyme-inhibitor complex in situ by X-ray crystallography. We have also reported the three-dimensional structure of HNE, the actual target enzyme of interest, but only in a pre-formed inhibited complex with a peptide chloromethyl ketone (Navia et. al., 1989). Unfortunately, uncomplexed HNE crystals are inadequate for high resolution X-ray structure determination (Williams et. al., 1987). However, a direct comparison of

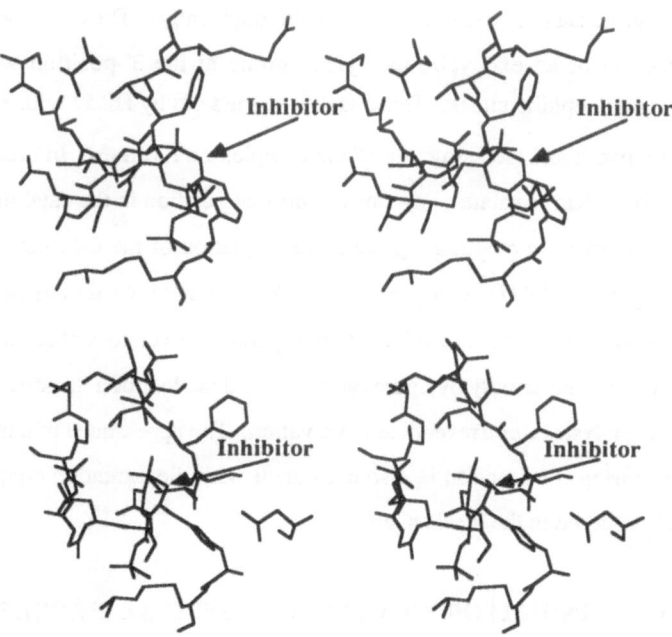

Figure 3. Stereo diagram of the active site of porcine pancreatic elastase showing the inhibitor L-647,957 bound. The two views are rotated 45° from each other. Details of the structure determination are given in Navia *et. al.*, 1987.

two enzyme structures suggests that what we have learned from the PPE - inhibitor complex studies in solution and in the crystalline state can be directly applied to the corresponding HNE complexes (Navia *et. al.*, 1989).

In our structure report, we also proposed a comprehensive mechanism of action, based on the observed PPE - L-647,957 complex. This mechanism appears to hold for HNE inhibition as well (Navia *et. al.*, 1987, 1989). Intrinsic binding of L-647,957 by both PPE and HNE is only modest (K_i = 0.3 and 0.6 μM respectively); as with the ß-lactamases, potent inhibition of the elastases requires a chemical reaction between inhibitor and enzyme. The first step involves acylation of the ß-lactam ring, mediated, in this case, by Ser 195 (k_2 = 0.02 sec^{-1} and 0.01 sec^{-1} for PPE and HNE respectively).

Acylation is followed by the expulsion of chloride from the 7-position of L-647,957, and of the acetate leaving group from the 3-position. (The exact order in which the reactions take place cannot be determined crystallographically). These reactions lead to the production of an exocyclic methylene group at the 3'-position which is susceptible to nucleophilic attack. This attack is carried out by His57 with a $t_{1/2}$ of approximately two hours (see structure of the complex on figure 3). In order to do this, residue His57 had to rotate away from its normal position in the catalytic triad, where it was involved in potentiating the initial acylation of the ß-lactam ring by Ser195 (see, *eg.* Kraut, 1977). Our proposed mechanism of action is consistent with the solution biochemistry of the reaction, including the observed loss of sensitivity to regeneration of enzyme activity by hydroxyl amine, and the loss of a histidine residue by amino acid analysis after irreversible inactivation. The appearance of a transient yellow color during the reaction is also consistent with the extended conjugated intermediates proposed in the mechanism.

ELASTASE - INHIBITOR COMPLEXES, SPECIAL PROBLEMS

Given that serine protease inhibition proceeds via an acyl enzyme intermediate (Kraut, 1977, Powers *et. al.*, 1986), it is not surprising that all known potent elastase inhibitors undergo some degree of chemical modification in their interaction with these enzymes (see *eg.* Table 1 of Copp *et. al.,* 1987). More often than not, in the case of the ß-lactam elastase inhibitors, the nature of the product (or products) of these reactions is unknown (or unknowable) prior to the solution of the crystal structure of the complex. At times, inhibitors undergo a whole series of reactions before arriving at an end state, where none of the reaction steps necessarily go to completion. Thus, electron density maps of these complexes are often uninterpretable in terms of a single inhibitory species. Compounds that are effectively irreversible in the time frame of a standard biochemical assay nonetheless often regenerate free enzyme (or worse, partial products of degeneration) in the much longer time frame of a crystallographic data collection experiment.

Standard crystallographic practice often creates problems too. For example, the addition of azide to control microbial growth in crystallization experiments introduces a powerful nucleophile into the system that is capable of deacylating inhibitors and regenerating free enzyme. In trying to form inhibitor complexes with PPE, the sulfate used to crystallize the enzyme can be a problem. In the native structure, a sulfate ion is located in the active site pocket of the enzyme, and appears to bind tightly enough to prevent binding by ß-lactams. As such, we have had to form our crystalline PPE - inhibitor complexes in crystals that were back-soaked out of sulfate solutions into sulfate-free 35% poly-ethylene glycol 6000. Finally, most of the ß-lactam elastase inhibitors we have examined are only modestly soluble in water, so that it has been necessary to use dimethyl sulfoxide (DMSO) solutions to obtain inhibitor concentrations appropriate to the very high protein concentrations used in these experiments. In many cases, the combination of backsoaking and exposure to DMSO is too much of a stress, destroying the crystals before data collection can be carried out.

Once data is collected and a structure is solved, the experimental outcome is a contour map of electron density, to which some kind of molecular interpretation must be attached. This is complicated, as was mentioned above, by the fact that the identity of the final enzymatically modified species is unknown in most instances. Our approach to this problem has been to consider all possibly feasible mechanisms for a reaction, and to construct on the computer corresponding models of modified inhibitors. In the structure determination of the PPE complex with L-647,957, construction of these models was aided by our knowledge of the small molecule crystal structure of L-647,256 (3-methyl-7-alpha-methoxy-3-cephem-4-carboxylate-1,1-dioxide tert-butyl ester), an analog of L-647,957 (see figure 4). The structure of L-647,256 was specifically solved for this study.

In the structure of the PPE complex with L-647,957, it was immediately clear that electron density flowed strongly and continuously from Ser195 into the inhibitor, so that all models considered incorporated this feature. Qualitative inspection of electron density maps also allowed some mechanisms to be discarded outright, *eg.* His 57 nucleophilic attack at C-6. In most other cases, models had to be systematically

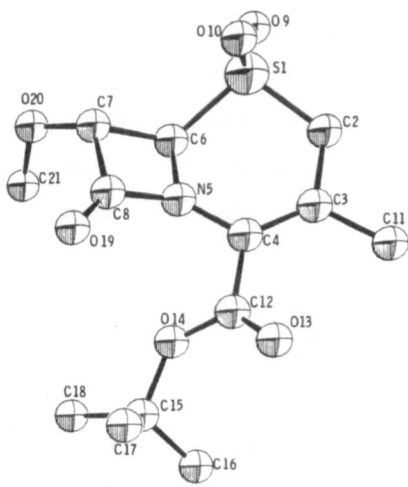

Figure 4 Computer-generated perspective drawing of the solved structure of L-647,256 (3-methyl-7-alpha-methoxy-3-cephem-4-carboxylate-1,1-dioxide tert butyl ester), a close analog of L-647,957 (see figure 2 above). An X-ray structure of a cephalosporin sulfone was not available for use in interpreting the experimentally derived difference electron density of the complex of porcine pancreatic elastase with L-647,957. Attempts to crystallize L-647,957 were unsuccessful, although L-647,256 did crystallize from methanol. Data from crystals of L-647,256 were collected on an Enraf-Nonius CAD-4 diffractometer using monochromated copper radiation. Preliminary diffraction experiments indicated that the space group symmetry was $P2_12_12_1$ with a=14.295(2)Å, b=6.006(1)Å, and c=18.317(2)Å for Z=4. Of the 1252 reflections collected, 1066 were observed (I>3*sigma(I)). The structure was solved using the program MULTAN, difference Fourier analysis, and full matrix least-squares refinement. Hydrogens were assigned isotropic temperature factors corresponding to their attached atoms. The function minimized was $\Sigma w(|F_o|-|F_c|)^2$ with $w=1/(sigma(F_o))^2$ to give an unweighted residual of 0.039. Tables containing the crystallographic coordinates, bond distances, and bond angles have been deposited with the Cambridge Crystallographic Data Center, Cambridge, England. The program ORTEP (Johnson, 1970) was used to plot the structure of L-647,256 from the final X-ray coordinates. The model building program MOLEDIT (Gund et. al., 1980) was used to construct a series of possible enzymatic modifications of L-647,957 based on the structure of L-647,256, which models were subsequently fitted to the fullest extent possible into the experimental difference electron density.

examined graphically with respect to their ability to account for the electron density observed.

4-CARBOXY ANALOGS ARE INEFFECTIVE AS ELASTASE INHIBITORS

Yet to be explained is the relative ineffectiveness of ß-lactamase-inhibitor-like, analogs *vs.* the elastases. This can be rationalized in part by the different sterochemistry and length of substituents found at the 7-position in the cephalosporins (corresponding to the 6-position in the penicillins). In the ß-lactam antibiotics, these substituentts are thought to form part of a D-Ala-D-Ala peptide mimic, corresponding to one of the components of bacterial cell wall synthesis (Waxman *et. al.*, 1983; Lamotte-Brasseur, *et. al.*, 1984). By such a mimic, the ß-lactam antibiotics are bound to, and ultimately inactivate the cell wall synthetic enzymes. In the elastase inhibitors, the 7-position mimics the P_1 side chain of substrate, as shown in figures 3 and 5, and as inferred from structure activity relationships vs. both PPE and HNE (Doherty *et. al,* 1986).

At the (cephalosporin) 4-position the substitution effects are more subtle. Even though all ß-lactam antibiotics possess a negatively charged group (usually a carboxyl) at that position, such a substitution in the ß-lactam elastase inhibitors makes them inactive (Doherty *et. al.*, 1986). Initially, we felt that the presence of Asp60 near His57 in the sequence of PPE might be involved in an electrostatic repulsion of the negative group at the 4-position. The structure of the PPE complex with L-647,957, as shown in figure 3, indicates that this is not the case.

Figure 5 shows a model of unreacted L-647,957 in the active site of native uncomplexed PPE, where the 4-position tert-butyl ester has been changed into a carboxyl and energy minimized. His57 is seen to be in close proximity to the 4-position carboxyl of the modelled "inhibitor". In this situation, one can speculate that the nearby negative charge might pull the histidine ring physically out of its functional position in the catalytic triad, or that it might significantly modify that residue's electronic properties, or both. With such an altered His57 residue, neither PPE or HNE would be able to function as an enzyme, so that an initial acylation reaction would not to take place.

38

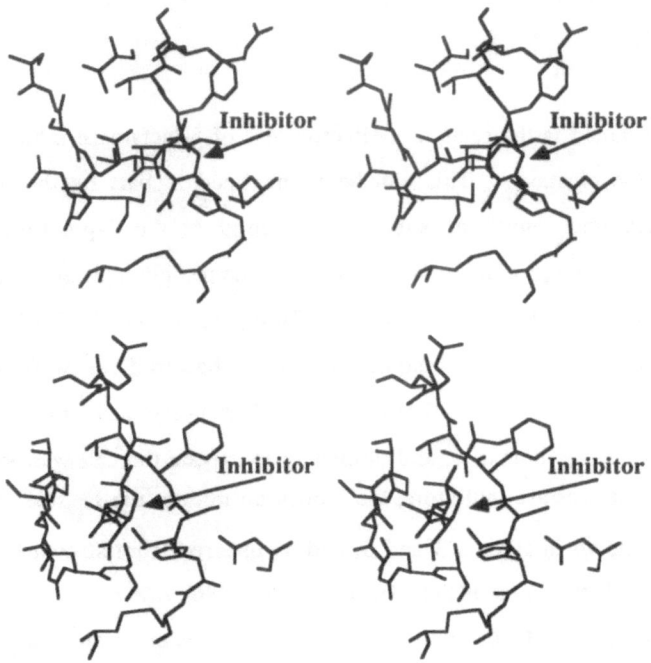

Figure 5. Structure of the active site of native porcine pancreatic elastase with a model of the 4-position free carboxyl analog of L-647,957 superimposed. The model was constructed by modifying the coordinates of the structure of the PPE complex with L-647,957 (Navia *et. al.*, 1987), using the QUANTA program system (Polygen Corp; Waltham, MA). The resulting structure was minimized using the CHARMm option within QUANTA. The negatively-charged 4'-carboxyl group is seen to be in close proximity to His57. One can speculate that the electronic properties of His 57 might be altered by this interaction, or that the residue might be pulled out of position in the catalytic triad. The two sterographic views shown are rotated by 45° from each other.

In retrospect, the characterization of the structure of the PPE complex with L-647,957 turned out to be relatively straight forward, mainly because the irreversible inhibition brought about by that compound leads to one species in the active site of the enzyme. The more interesting and therapeutically relevant inhibitors of elastase, however, will be those which are reversible, with a regeneration time compatible with once-a-day dosing. Unfortunately, conventional X-ray diffraction methods are simply too slow for the collection of suitable data in such cases, even with the use of modern area detectors. Such problems are ideally suited, however, for the ultrarapid data collection schemes based on the Laue method (Hajdu et. al., 1987) using intense beams of multiple wavelength X-rays produced at a synchrotron X-ray source. Turn over of these essentially "slow substrates" should be slow enough to allow for the uniform diffusion and reaction of inhibitor throughout the body of an enzyme-inhibitor complex crystal. Alternatively, cryogenic methods might allow one to trap the inhibited enzyme species long enough for data collection to take place.

In conclusion, inhibition of enzymes like the elastases, which incorporate an acylation step in their mechanism of action, appears to require a corresponding covalent link with inhibitor, if these are to be in any way potent enough to be pharmacologically useful (see eg. Table 1 of Copp *et. al.,* 1987). These chemical reactions lead to problems of interpretation which may not be as serious in the examination of inhibitor complexes with enzymes that do not go through an intermediate covalently bonded stage in their mechanism (*eg.* the aspartyl proteases and Zn^{++} metalloenzymes).

ACKNOWLEDGEMENTS

We thank Dr. D. Underwood for guidance in using the QUANTA system of programs, and Drs. J. B. Doherty, T.-Y. Lin, and their colleagues in the elastase inhibitor program for their continuing assistance and support.

REFERENCES

Bush, K. 1988. ß-lactamase Inhibitors from Laboratory to Clinic. Clin. Microbiol. Rev. 1:109-123.

Carp, H., Miller, F., Hoidal, J.R., and Janoff, A. 1982. Potential Mechanism of Emphysema: Alpha$_1$-Proteinase Inhibitor Recovered from Lungs of Cigarette Smokers Contains Oxidized Methionine and Has Decreased Elastase Inhibitory Capacity. Proc. Natl. Acad. Sci. USA 79:2041-2045.

Carrell, R.W., Jeppson, J.-O., Laurell, C.-B., Brennan, S.O., Owen, M.C., Vaughan, L., and Boswell, D.R. 1982. Structure and Variation of Human Alpha$_1$-Antitrypsin. Nature 298:329-334.

Carrell, R.W. 1986. Alpha$_1$-Antitrypsin: Molecular Pathology, Leukocytes, and Tissue Damage. J. Clin Invest. 78:1427-1431.

Clark, J.M., Vaughan, D.W., Aiken, B.M., and Kagan, H.M. 1980. Elastase-Like Enzymes in Human Neutrophils Localized by Ultrastructural Cytochemistry. J. Cell Biol. 84:102-119.

Cochrane, C.G., Spragg, R.G., Revak, S.D., Cohen, A.B., and McGuire, W.W. 1983. The Presence of Neutrophil Elastase and Evidence of Oxidation Activity in Bronchoalveolar Lavage Fluid of Patients with Adult Respiratory Distress Syndrome. Am. Rev. Respir. Dis. 127:S25-S27.

Copp, L.J., Krantz, A., and Spencer, R.W. 1987. Kinetics and Mechanism of Human Leukocyte Elastase Inactivation by Ynenol Lactones. Biochemistry 26:169-178.

Doherty, J.B., Ashe, B.M., Argenbright, L.W., Barker, P.L., Bonney, R.J., Chandler, G.O., Dahlgren, M.E., Dorn, C.P. Jr., Finke, P.E., Firestone, R.A., Fletcher, D., Hagmann, W.K., Mumford, R., O'Grady, L., Maycock, A.L., Pisano, J.M., Shah, S.K., Thompson, K.R., and Zimmerman, M. 1986. Cephalosporin Antibiotics Can be Modified to Inhibit Elastase. Nature 322:192-194.

Gadek, J.E. and Crystal, R.G. 1983. Experience with Replacement Therapy in the Destructive Lung Disease Associated with Severe Alpha$_1$-Antitrypsin Deficiency. Am. Rev. Respir. Dis. 127:S45-S46.

Gund, P., Andose, J.D., Rhodes, J.B., and Smith, G.M. 1980. Three-Dimensional Molecular Modeling and Drug Design. Science 208:1425-1431.

Hajdu, J., Machin, P.A., Campbell, J.W., Greenhough, T.J., Clifton, I.J., Zurek, S., Grover, S., Johnson, L.N., and Elder, M. 1987. Millisecond X-ray Diffraction and the First Electron Density Map from Laue Photographs of a Protein Crystal. Nature 329:178-181.

Herzberg, O., and Moult, J. 1987. Bacterial Resistance to ß-Lactam Antibiotics: Crystal Structure of ß-Lactamase from Staphylococcus aureus PC1 at 2.5Å Resolution. Science 236:694-701.

Janoff, A. 1978. Granulocyte Elastase: Role in Arthritis and in Pulmonary Emphysema. In: "Neutral Proteases of Human Polymorphonuclear Leukocytes (K. Havemann and A. Janoff, eds.) Urban & Schwartzenberg, Baltimore-Munich.

Janoff, A. 1985. Elastase in Tissue Injury. Ann. Rev. Med. 36:207-216.

Johnson, C.K. 1970. ORTEP-II program, Oak Ridge National Laboratory, Oak Ridge, TN.

Knowles, J.R. 1985. Penicillin Resistance: The Chemistry of ß-Lactamase Inhibition. Acc. Chem. Res. 18:97-104.

Kraut, J. 1977. Serine Proteases: Structure and Mechanism of Catalysis. Ann Rev. Biochem. 46:331-358.

Lamotte-Brasseur, J., Dive, G., and Ghuysen, J.-M. 1984. On the Structural Analogy Between D-alanyl-D-alanine Terminated Peptides and ß-Lactam Antibiotics. Eur. J. Med. Chem. - Chim. Ther. 19:319-330.

Navia, M.A., Springer, J.P., Lin, T.-Y., Williams, H.R., Firestone, R.A., Pisano, J.M., Doherty, J.B, Finke, P.E., and Hoogsteen, K. 1987. Crystallographic Study of a ß-Lactam Inhibitor Complex with Elastase at 1.84Å Resolution. Nature 327:79-82.

Navia, M.A., McKeever, B.M., Springer, J.P., Lin, T.-Y., Williams, H.R., Fluder, E.M., Dorn, C.P. Jr., and Hoogsteen, K. 1989. Structure of Human Neutrophil Elastase in Complex with a Peptide Chloromethyl Ketone Inhibitor at 1.84Å Resolution. Proc. Natl. Acad. Sci. USA 86:7-11.

Parmley, R.T., Doran, T., Boyd, R.L., and Gilbert, C. 1986. Unmasking and Redistribution of Lysosomal Sulfated Glycoconjugates in Phagocytic Polymorphonuclear Leukocytes. J. Histochem. Cytochem. 34:1701-1707.

Powers, J.C., and Harper, J.W. 1986. Inhibitors of Serine Proteases. In: "Proteinase Inhibitors" (A. Barrett and G. Salvesen, eds) Elsevier, New York.

Ranga, V., Kleinerman, J., Ip, M.P.C., Sorensen, J., and Powers, J.C. 1981. Effects of Oligopeptide Chloromethylketone Administered after Elastase: Renal Toxicity and Lack of Prevention of Experimental Emphysema. Am. Rev. Respir. Dis. 124:613-618.

Robbins, S.L., Cotran, R.S., and Kumar, V. 1984. Emphysema. In: "Pathologic Basis of Disease". W.B. Saunders Co. Philadelphia.

Sawyer, L., Shotton, D.M., Campbell, J.W., Wendell, P.L., Muirhead, H., Watson, H.C., Diamond, R., and Ladner, R.C. 1978. The Atomic Structure of Crystalline Porcine Pancreatic Elastase at 2.5Å Resolution: Comparisons with the Structure of alpha-Chymotrypsin. J. Mol. Biol. 118:137-208.

Sinha, S., Watorek, W., Karr, S., Giles, J., Bode, W., and Travis, J. 1987. Primary Structure of Human Neutrophil Elastase. Proc. Natl. Acad. Sci. USA 84:2228-2232.

Waxman, D.J., and Strominger, J.L. 1983. Penicillin-Binding Proteins and the Mechanism of Action of ß-Lactam Antibiotics. Ann. Rev. Biochem. 52:825-869.

Williams, H.R., Lin, T.-Y., Navia, M.A., Springer, J.P., McKeever, B.M., Hoogsteen, K., and Dorn, C.P. Jr. 1987. Crystallization of Human Neutrophil Elastase. J. Biol. Chem. 262:17178-17181.

Design of Purine Nucleoside Phosphorylase Inhibitors Using X-Ray Crystallography

STEVEN E. EALICK, Y.S. BABU, S.V.L. NARAYANA, WILLIAM J. COOK, and CHARLES E. BUGG

Introduction

Purine nucleoside phosphorylase (PNP) catalyzes the reversible phosphorolysis of ribonucleosides and 2'-deoxyribonucleosides to the free base and ribose-1-phosphate (1-4). Although equilibrium favors the synthesis of nucleosides, the biological function of PNP is to liberate purine bases through the purine salvage pathway. PNP has been isolated from a variety of sources including human erythrocytes, bovine spleen, chicken liver, rabbit erythrocytes, E. coli and avian malarial parasite (5-15). In general, it appears that the mammalian enzyme is a trimer of identical subunits (16-20) while the bacterial enzyme is a hexamer of identical subunits (21). Both the human erythrocytic PNP and the E. coli PNP have been crystallized (22, 23).

The human erythrocytic enzyme is the most thoroughly studied PNP. The enzyme has been cloned (24) and the cDNA derived amino acid sequence (25) has been published (Figure 1). The molecular weight of human PNP is about 97,000 daltons and the subunits consists of identical 289 amino acid polypeptide chains. Human PNP is specific for analogs of guanosine and inosine although detectable activity has been reported for the cleavage of adenosine (26). The general requirement appears to be a hydrogen bond acceptor at C(6) and a hydrogen bond donor at N(1) of the purine base. Purines such as guanine, which have an amino group at C(2), show enhanced binding. The enzyme will accept ribonucleosides and 2'-deoxyribonucleosides equally well, but will not accept purine arabinosides as substrates (27).

PNP isolated from E. coli has a molecular weight of about 150,000 daltons with a subunit weight of about 25,000 daltons (23). As opposed to the human enzyme, adenosine is a good substrate for E. coli PNP. In addition, the E. coli enzyme will also accept arabinosides as substrates (27). Although the E. coli enzyme has much broader specificity, the human enzyme is much more efficient in cleaving the narrower range of substrates it accepts (28-30). These fundamental differences between the human and E. coli PNPs have led us to initiate crystallographic studies of the bacterial enzyme.

Initial interest in the human enzyme came from the observation that PNP cleaves some purine nucleoside analogs of potential chemotherapeutic value (31-35). Because PNP is present at high activity, nucleoside analogs, such as 6-mercaptopurine riboside and 6-thioguanosine are rapidly cleaved to the less effective base before they can reach the desired site of action. More recently,

43

44

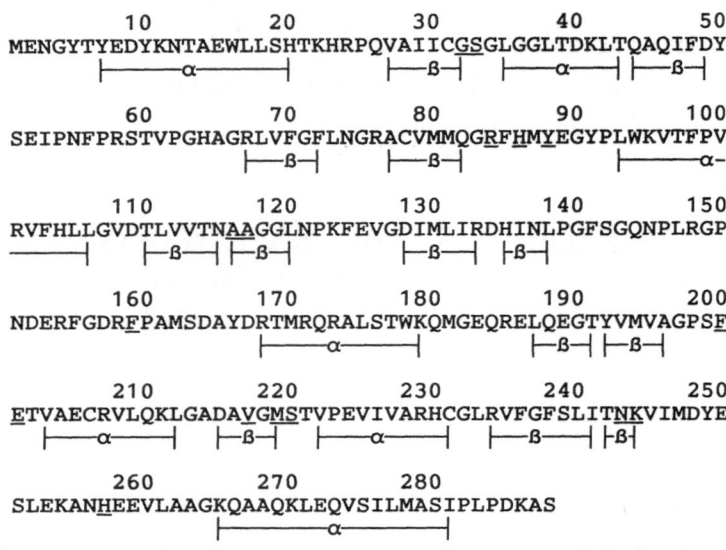

Figure 1. Amino acid sequence for human purine nucleoside phosphorylase

it has been suggested that degradation by PNP of the promising anti-AIDS drug, dideoxyinosine, might be responsible for its extremely short half-life in humans.

PNP has also been shown to play a role in immune function. Absence of PNP has been associated with a severe T-cell immune deficiency while B-cell function remains normal (36, 37). Several cases have been reported of patients who have a genetic defect which leads to little or no PNP activity. These extremely rare patients display highly suppressed or no T-cell immune response. It has been proposed that in T-cells, most of the 2'-deoxyguanosine is converted to 2' deoxyguanosine triphosphate in the absence of PNP. This accumulation of dGTP leads to feedback inhibition of ribonucleotide reductase and T-cell death (38).

Because of the key role of PNP in purine nucleoside metabolism and the role of immunodevelopment, a number of clinical applications for PNP inhibitors have been suggested. PNP inhibitors might potentiate the action of certain purine nucleoside analogs, such as dideoxyinosine or 2' deoxythioguanosine, by protecting them from phosphorolysis by PNP. Since PNP inhibitors should lead to selective suppression of the T-cell immune response, PNP inhibitors might be used to selectively kill T-cells in T-cell leukemia, to treat autoimmune disease such as rheumatoid arthritis or to suppress the host-vs-graft response in organ transplant patients. The literature concerning PNP as a target for chemotherapy has been recently reviewed (39).

Despite the potential therapeutic applications of PNP inhibitors, no really good membrane permeable inhibitors have been identified. To date the best inhibitors are in the 10^{-6} to 10^{-7} range and include 8-aminoguanine (40-41), 5-iodoformycin B (4), 5'-iodo-9-deazainosine (42), 8-amino-9-benzylguanine (43). Acyclovir diphosphate has a Ki of around 10^{-9} when measured in low phosphate but has no potential as a therapeutic agent because it would not cross the membrane and is a substrate for cellular kinases (44, 45). Because traditional trial and error approaches had not yet been successful in producing a good PNP inhibitor, we decided to use X-ray crystallography in an attempt to design tight binding inhibitors of human PNP.

Crytstallization

Crystals of PNP from human erythrocytes had been reported several years ago (46). However, these crystals were very small and unusable for X-ray crystallography. Purified PNP was obtained from Drs. Johanna Stoeckler and Robert Parks at Brown University. The previously reported needles were obtained from ammonium sulfate around pH = 7. At lower pH (5.2-5.4) rhombohedral crystals grew up to 0.6mm on an edge. These crystals belong to space group R32 and have unit cell dimensions of a = 143Å and c = 165Å. The crystals have one subunit per asymmetric unit and about 76% solvent. The crystals are moderately stable in an X-ray beam and diffract to about 2.7Å resolution.

PNP purified from human erthrythrocytes displays isoelectric heterogeneity (47). It has been proposed that the heterogeneity arises from post-translational modification such as deamidation of glutamine to glutamate and asparagine to aspartate. Recently we obtained some wild type recombinant human PNP from Drs. Steve Williams and David Martin of Genentech. Although some large crystals were obtained at pH = 5.4, the crystals grew slowly and infrequently (Figure 2). Around pH = 7 large pointed plates appeared with a longest dimension of about 0.6mm (Figure 3). Because

Figure 2.   Crystals of recombinant PNP grown from ammonium sulfate at pH = 5.4.

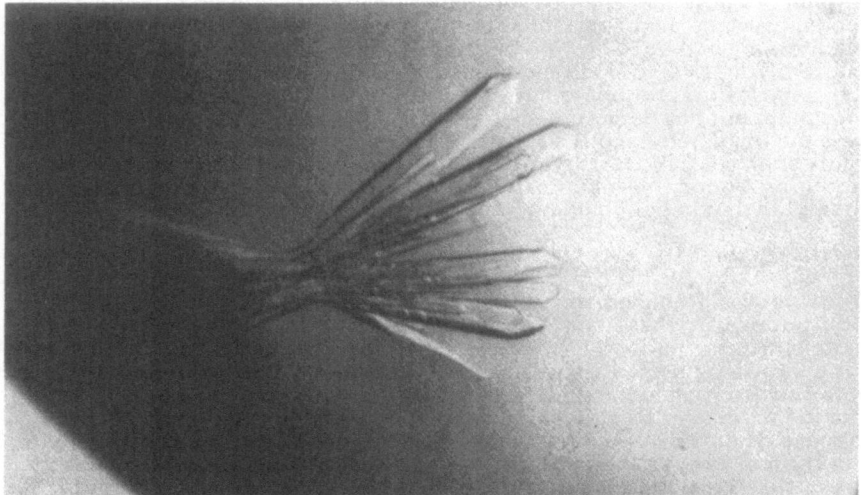

Figure 3. Crystals of recombinant PNP grown from ammonium sulfate at pH-7.0.

the crystals obtained earlier at pH = 7 were so small the morphology for these crystals could not be determined. However, it is likely that the improved crystal size for the recombinant PNP is due to isoelectric homogeneity. Preliminary X-ray photographs showed that the pH = 7 crystals diffracted to about the same resolution as the rhombohedral crystals and further X-ray studies have not been pursued to date.

Three-Dimensional Structure

The structure of human PNP has now been determined using single crystal X-ray diffraction techniques (48). Because the crystals were only moderately stable in an X-ray beam and diffracted to only 2.7Å resolution, we used synchrotron radiation and oscillation photography for the structure determination. Data for native crystals and four heavy-atom derivatives was collected at the Synchrotron Radiation Source (SRS), Daresbury, England. In addition, the data for a number of PNP/inhibitor and PNP/substrate complexes was collected at the SRS. The structure of PNP was determined at 3.2Å resolution using multiple-isomorphous-replacement techniques and refined using restrained least-squares techniques (49).

Figure 4 shows a ribbon drawing of a PNP trimer and Figure 5 shows a representation of a single subunit. PNP is an alpha/beta protein consisting of two beta sheets and seven alpha helices per subunit. In the center of the subunit is an eight stranded mixed beta sheet. A smaller five stranded beta sheet packs against the larger sheet to form a distorted beta barrel (Figure 6). This core beta structure is flanked by seven alpha helices.

The active site of PNP, which lies near the subunit interface, was characterized using several PNP complexes. Figure 7 shows the electron density for 9-benzyl-guanine as obtained from a cross difference Fourier map. Analysis of the active site suggested that several residues are important for binding and/or catalysis. These are shown schematically in Figure 8. It appears that specificity for inosine and guanosine arises primarily from Glu 201, Asn 243, and Lys 244. Both Asn 243 and Lys 244 donate hydrogen bonds to the carbonyl oxygen attached to C(6). Glu 201 accepts a hydrogen bond from N(1) and in the case of guanine also the amino group at C(2). The presence of this additional hydrogen bond may be responsible for the increased affinity for guanine relative to hypoxanthine. The remainder of the purine binding site consists primarily of hydrophobic residues including Ala 116, Phe 200, Val 217, and Met 219.

The phosphate binding site consists of Ser 33, Arg 84, His 86, and Thr 220. The amide hydrogen atoms from Ser 33 and Ala 116 are also positioned to participate in hydrogen bonding with the phosphate ion. In addition, it appears that a hydrogen bond may exist between O(3') of the ribose and the phosphate. In the crystal structure, the phosphate binding site appears to be occupied by a sulfate ion due to the use of ammonium sulfate in the crystallization procedure.

See color insert:

Figure 4. Ribbon drawing of a PNP trimer. Each subunit is shown in a different color. One of three identical active sites is shown with nucleoside and phosphate bound.

See color insert:

Figure 5. Ribbon drawing of a PNP subunit. The color coding is by amino acid type. Green = hydrophobic; Blue - basic; Red - acidic; Orange - alcohols; Purple = Gln and Asn; Light Blue = Gly and Pro; Yellow = Met.

See color insert:

Figure 6. Stereoview of the beta sheet structure for PNP.

See color insert:

Figure 7. Difference electron density for 9-benzylguanine.

Figure 8 .   Arrangement of active site residues.

The ribose group of purine nucelosides appears to be involved in few specific interactions. In addition to the proposed hydrogen bond between O(3') and phosphate, there is a hydrogen bond between O(3') and Tyr 88. No hydrogen bond is formed with O(2') and the nearest amino acid is Met 219. This is consistent with the observation that binding of substrate is not sensitive to the presence or absence of O(2'). The absence of specific interactions with ribose is consistent with the nature of the reaction that the enzyme catalyzes. The reaction is a transfer reaction in which the ribose is reversibly transferred from purine base to phosphate. The lack of tight binding of ribose facilitates the transfer. In fact, the enzyme has no measurable affinity for ribose itself. The main purpose of the ribose pocket is to orient the nuceloside for nucleophilic attack. The pocket is formed from several aromatic hydrophobic residues including Tyr 88, Phe 159 (from an adjacent subunit), Phe 200, and His 257. Since the ribose ring has both a hydrophobic and a hydrophilic side, the sugar orients such that the C-H bonds point towards the hydrophobic and the hydroxyl groups point towards the phosphate binding site. This arrangement leads to the geometry required for nucleophilic attack at C(1') with inversion of the asymmetric center (Figure 9).

Recently, the sequence for bovine PNP was provided to us by Drs. Williams and Martin of Genentech. The bovine and human enzymes show considerable similarity of specificity and kinetic properties. Comparison of the sequences of bovine and human enzymes indicated 87% exact homology. Furthermore, every amino acid believed to be important for binding or catalysis was conserved. In general, the unconserved residues lie on the surface of the protein (Figure 10).

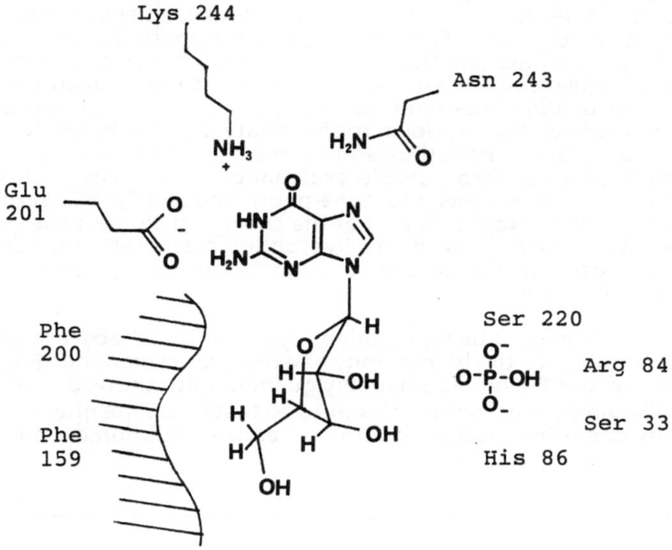

Figure 9.   Arrangement of reactants for the phosphorolytic reaction.

The sequence of PNP from a patient with T-cell immune deficiency has also been reported (50). It was found that amino acid 89, which is a glutamate in the wild type PNP, had become a lysine as the result of a single base change. Although this residue is not directly involved in substrate binding, its role can be proposed. His 86 is located between Glu 89 and the phosphate ion (Figure 11). It is possible that a proton is transferred from the phosphate to His 86 to provide a better nucleophile for the phosphorolysis. The presence of Glu 89 would stabilize the protonated form of His 86. If Glu 89 were replaced by a positively charged lysine, protonation of His 86 would be unlikely.

Inhibitor Binding

In order to understand inhibitor binding PNP complexes with 8-aminoguanine, acyclovir diphosphate and 8-amino-9-benzylguanine have been examined using X-ray diffraction techniques. Structure activity data show that an amino group at C(8) of the purine ring usually enhances binding (40, 41). This can be explained on the basis that the amino group is in a position to donate hydrogen bonds to the main chain carbonyl of Ala 116 and the side chain of Thr 242 (Figure 12). By satisfying both hydrogen bond donors the complex maintains a low energy and also takes advantage of entropic effects relative to solvated 8-aminoguanine.

Acyclovir diphosphate is particularly interesting because the Ki is much better at low phosphate than at high phosphate and both the corresponding monophosphate and the triphosphate are significantly poorer inhibitors (44, 45). This data suggested that acyclovir diphosphate occupies both the purine and phosphate binding site. Furthermore, it was suggested that the spacing between the purine base and the terminal phosphate was optimum for the diphosphate. Analysis of the X-ray structure showed that indeed the purine and phosphate binding sites were occupied (Figure 13). Simple docking experiments showed that acyclovir triphosphate was probably too large. However, docking experiment showed that the monophosphate could easily span the two subsites. Two possible explanations exist. First, a hydrogen bond can be proposed between the alpha-phosphate and Tyr 88. Secondly, it was found that the hydrophobic part of the chain that connect the base and phosphate interacts closely with the hydrophobic aromatic pocket which orients the ribose. In the case of the monophosphate, neither of the interactions could occur.

In the case of 8-amino-9-benzylguanine, hydrophobic effects are probably entirely responsible for the high affinity. Two similar compounds 8-amino-9-phenylguanine and 8-amino-9-phenethylguanine both showed much larger Ki values (43). Analysis of the complex showed that when the phenyl group is separated by one (-CH2-) spacer, it forms a nearly ideal interaction with Phe

---

See color insert:

Figure 10. Ribbon drawing comparing the amino acid sequences of human and bovine PNPs. Green represents identical residues, yellow represents conservative substitutions and red represents non-conservative substitutions.

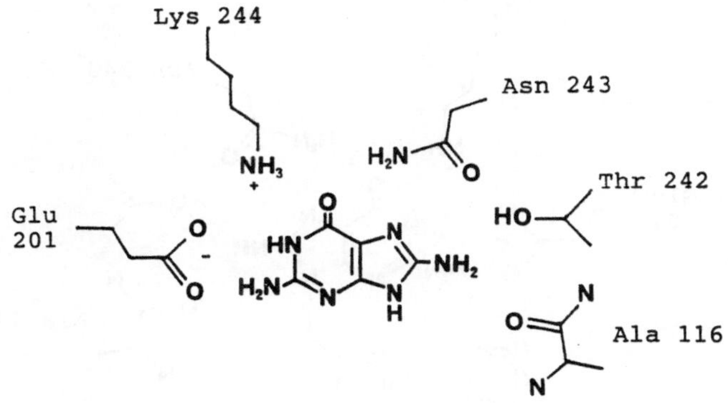

Figure 11. Hydrogen bonding arrangement for 8-amino analogs.

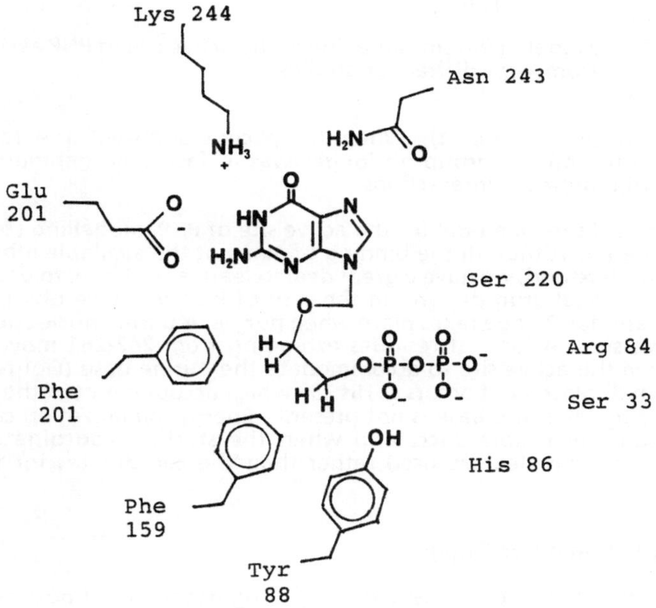

Figure 12. Model of acyclovir diphosphate in the active site of PNP as determined from X-ray diffraction studies.

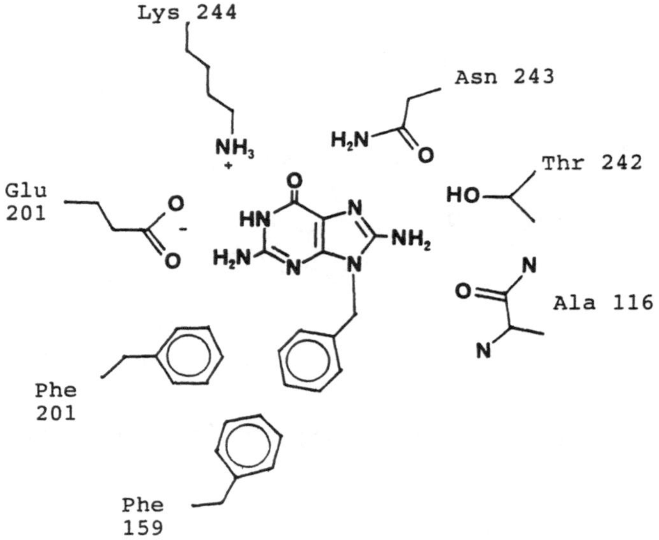

Figure 13. Model of 9-benzylguanine in the active site of PNP as determined from X-ray diffraction studies.

159 and Phe 200 (Figure 11). When the spacer was absent or extended to (-CH$_2$-CH$_2$-) the phenyl group no longer was in favorable geometry to form hydrophobic aromatic interactions.

Having established a model for the active site of PNP, modelling studies have been initiated. Although the binding of many of the available inhibitors can now be understood, we have a great deal to learn about how to use structural data for rational drug design. In the case of PNP, we have observed that a conformational change takes place when purine or purine nucleoside binds to the active site. A loop of residues extending from 242-261 moves out and away from the active site to accommodate the purine base (Figure 14). The maximum displacement occurs at His 257 which occupies part of the purine binding site when the base is not present. Energy minimization calculations have been much more successful when the starting coordinates for the PNP/guanine complex are used rather than the coordinates for the native enzyme.

Strategy for Inhibition Design

We have decided to treat the process of drug design as a bootstrap process. The overall strategy is shown in Figure 15. A suggested target is first modelled in the active site of PNP using energy minimization techniques. Promising compounds are synthesized and assayed for *in vitro* activity using a IC$_{50}$ value as an indicator. The PNP/inhibitor complex is then prepared and analyzed using X-ray crystallography. The results are compared with the predictions based on modelling and then modifications are suggested. The loop is then repeated until the lead compound has been optimized. Although this process still remains trial-and-error to some extent, we are able to identify promising leads with much reliability. Using this strategy, we have obtained novel tight binding inhibitors which will be discussed in future publications.

See color insert:

Figure 14. Conformational changes in PNP upon binding of guanine. Line segments are drawn between the position of the atom in the native structure and the atom in the guanine complex. Colors range from red for the longest lines to blue for the shortest lines.

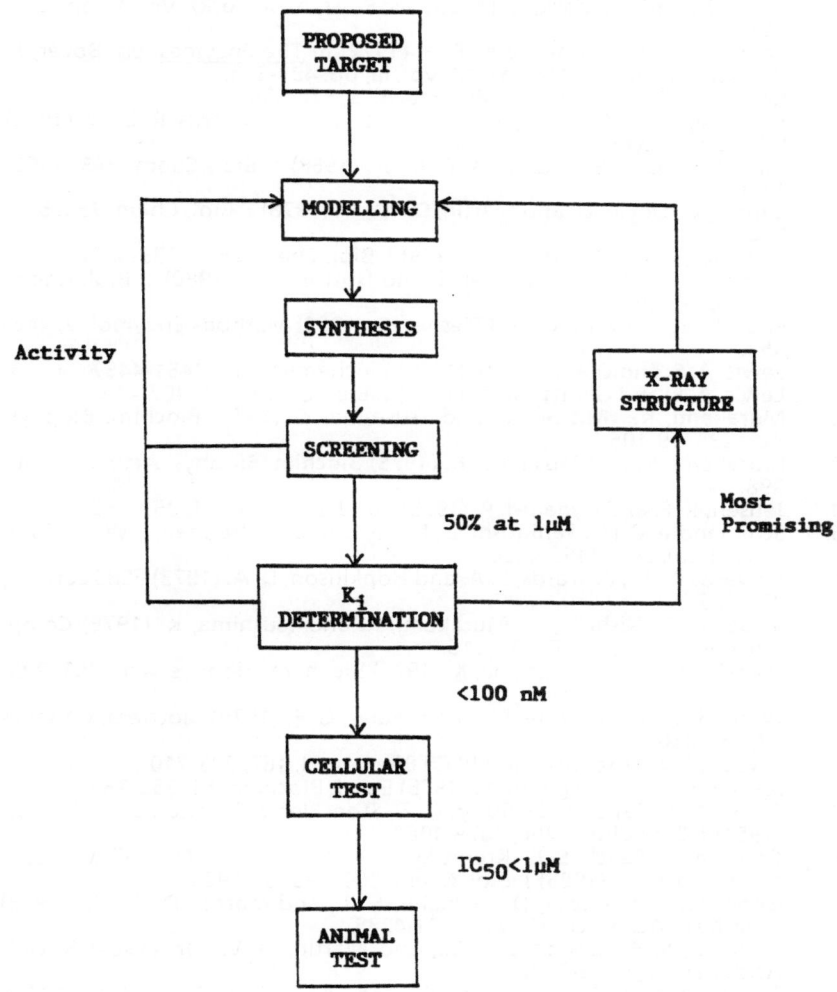

Figure 15. Flow chart describing the strategy for rational design of PNP inhibitors using X-ray crystallography.

54

Acknowledgments

This research was supported by grant GM-38823 from the National Institutes of Health, grant CH-213 from the American Cancer Society and by the Leukemia Society of America. The authors gratefully acknowledge the assistance of Mrs. Cookie Woodruff in the preparation of this manuscript.

References

1. Friedkin, M. and Kalckar, H. M. (1961) in The Enzymes, eds. Boyer, P.D., Lardy, H. and Myrback, K. (Academic Press, New York), Vol. V, pp. 237-255.
2. Parks, R. E., Jr. and Agarwal, R. P. (1972) in The Enzymes, ed. Boyer, P. D., (Academic Press, New York), Vol. III, pp. 483-514.
3. Krenitsky, T. A. (1967) Mol. Pharmacol. 3, 526-536.
4. Stoeckler, J. D., Agarwal, R. P., Agarwal, K. C. and Parks, R. E., Jr. (1978) Methods Enzymol. 51, 530-538.
5. Kim, B. K., Cha, S. and Parks, R. E., Jr. (1968) J. Biol. Chem. 243, 1763-1770.
6. Zannis, V., Doyle, D. and Martin, D. W., Jr. (1978) J. Biol. Chem. 253, 504-510.
7. Ghangas, G. and Reem, G. H. (1979) J. Biol. Chem. 254, 4233-4237.
8. Wiginton, D. A., Coleman, M. S. and Hutton, J. J. (1980) J. Biol. Chem. 255, 6663-6669.
9. Price, V. E., Otey, M. C. and Plesner, P. (1955) Methods Enzymol. 2, 448-453.
10. Lewis, A. S. and Glantz, M. D. (1976) Biochemistry 15, 4451-4457.
11. Lewis, A. S. and Glantz, M. D. (1976) J. Biol. Chem. 251, 407-413.
12. Murakami, K., Mitsui, A., and Tsushima, K. (1971) Biochim. Biophys. Acts 235, 99-105.
13. Murakami, K. and Tsushima, K. (1975) Biochim. Biophys. Acta 384, 390-398.
14. Jensen, K. F. and Nygaard, P. (1975) Eur. J. Biochem. 51, 253-265.
15. Schimandle, C. M., Tanigoshi, L., Mole, L. A. and Sherman, I. W. (1985) J. Biol. Chem. 260, 4455-4460.
16. Edwards, Y. H., Edwards, P. A. and Hopkinson, D. A. (1973) FEBS Lett. 32, 235-237.
17. Ikezawa, Z., Nishino, T., Murakami, K. and Tsushima, K. (1978) Comp. Biochem. Physiol. 60B, 111-116.
18. Murakami, K. and Tsushima, K. (1976) Biochem. Biophys. Acta 453, 205-210.
19. Ward, R. D., McAndrew, B. J. and Wallis, G. P. (1979) Biochem. Genetics 17, 251-256.
20. Savage, B. and Spencer, N. (1977) Biochem. J. 167, 703-710.
21. Jensen, K. F. and Nygaard, P. (1975) Eur. J. Biochem. 51, 253-265.
22. Cook, W. J., Ealick, S. E., Bugg, C. E., Stoeckler, J. D., and Parks, R. E., Jr. (1981) J. Biol. Chem. 256, 4079-4080.
23. Cook, W. J., Ealick, S. E., Krenitsky, T. A., Stoeckler, J. D., Helliwell, J. R. and Bugg, C. E. (1985) J. Biol. Chem. 260, 12968-12969.
24. Goddard, J.M., Caput, D., Williams, S. R., and Martin, D. W., Jr. (1983) Proc. Natl. Acad. Sci. USA 80, 4281-4285.
25. Williams, S. R., Goddard, J. M., and Martin, D. W., Jr. (1984) Nucleic Acids Res. 12, 5779-5787.
26. Zimmermon, T. P., Gersten, N. B., Ross, A. F., and Miech, R. P. (1971), Can. J. Biochem. 49, 1050-1054.
27. Stoeckler, J. D., Cambor, C., and Parks, R. E., Jr. (1980) Biochemistry 19, 102-107.
28. Krenitsky, T. A. (1967) Mol. Pharmacol. 3, 526-536.

29. Krenitsky, T. A., Elion, G. B., Strelitz, R. A., and Hitchings, G. H. (1967) J. Biol. Chem. 242, 2675-2682.

30. Krenitsky, T. A., Elion, G. B., Henderson, A. M., and Hitchings, G. H. (1968) J. Biol. Chem. 243, 2876-2881.

31. Montgomery, J. A., Schabel, Jr., F. M., and Skipper, H. E. (1962) Cancer Res., 22, 50.

32. Rich, M. A., Perez, A. G., and Eitlinoff, M. L. (1962) Cancer Res. 22, 504.

33. Krokoff, I. H., Ellison, R. R. and Tan, C. T. C. (1961) Cancer Res. 21, 1015.

34. Skipper, H. E., Thomson, J. R., Hutchison, D. J., Schabel, Jr., F. M., and Johnson, Jr., J. J. (1967) Proc. Soc. Exptl. Biol. Med., 95, 135.

35. Sheen, M. R., Kim, B. K., and Parks, Jr., R. E. (1968) Mol. Pharmacol. 4, 293.

36. Giblett, E. R., Ammann, A. J., Wara, D. W., and Diamond, L. K. (1975) Lancet, 1, 1010-1013.

37. Stoop, J. W., Zegers, B. J. M., Henricks, G. F. M., Siegenbeek van Heukelom, L. H., Staal, G. E. J., de Bree, P. K., Wadman, S. K., and Ballieux, R. E. (1977) N. Engl. J. Med., 196, 651-655.

38. Osborne, W. R. A. and Barton, R. W. (1986) Immunology, 59, 63-67.

39. Stoeckler, J. D. (1984) in Developments in Cancer Chemotherapy, ed. Glazer, R. I. (CRC Press, Boca Raton), pp. 35-60.

40. Parks, R. E., Jr., Stoeckler, J. D., Cambor, D., Savarese, T. M., Crabtree, G. W. and Chu, S.-H. (1981) in Molecular Actions and Targets for Cancer Chemotherapeutic Agents, eds. Sartorelli, A., Lazo, J. S. and Bertino, J. R. (Academic Press, New York) pp. 229-252.

41. Kazmers, I. S., Mitchell, B. S., Dadonna, P. E., Wotring, L. L., Townsend, L. B., and Kelley, W. N. (1981) Science 214, 1137-1139.

42. Stoeckler, J. D., Ryden, J. B., Parks, R. E., Jr., Chu, M. Y., Lim, M. I., Ren, W. Y. and Klein, R. S. (1986) Cancer Res. 46, 1774-1778.

43. Shewach, D. S., Chern, J.-W., Pillote, K. E., Townsend, L. B. and Dadonna, P. E. (1986) Cancer Res. 46, 519-523.

44. Tuttle, J. V. and Krenitsky, T. A. (1984) J. Biol. Chem. 259, 4065-4069.

45. Stein, J.M., Stoeckler, J. D., Li, S. Y., Tolman, R. L., MacCoss, M., Chen, A., Karkas, J. D., Ashton, W. T., and Parks, R. E., Jr. (1987) Biochem. Pharmacol. 36, 1237-1244.

46. Agarwal, R. P. and Parks, R. E., Jr. (1969) J. Biol. Chem. 244, 644-647.

47. Agarwal, K. C., Agarwal, R. P., Stoeckler, J. D., and Parks, R. E., Jr. Purine Nucleoside Phosphorylase. Microhetereogeneity and Comparison of Kinetic Behavior of the Enzyme from Several Tissues and Species. Biochemistry 14, 79, 1975.

48. Ealick, S. E., Rule, S. A., Carter, D. C., Greenhough, T. J., Babu, Y. S., Cook, W. J., Habash, J., Helliwell, J. R., Stoeckler, J. D., Parks, R. E., Jr., Chen, S.-F., and Bugg, C. E. (1989) J. Biol. Chem. in press.

49. Henrickson, W. A. and Konnert, J. H., (1980) In Computing in Crystallography, Diamond, R., Ramaseshan, S. and Bentkatesan, K., (eds), Indian Academy of Sciences, Bangalore, 13.01-13.23.

50. Williams, S. R., Cockler, V., McIvor, R. S., and Martin, D. W., Jr. (1987) J. Biol. Chem., 262, 2332-2338.

The Potential Role of Solvation in the Dihydrofolate Reductase Species Selectivity of Trimethoprim

LEE F. KUYPER

Trimethoprim (1, TMP) is an important antibacterial agent (Finland et al., 1982) that exerts its activity through the inhibition of the enzyme dihydrofolate reductase (DHFR) (Hitchings, 1983). DHFR catalyzes the NADPH-dependent reduction of dihydrofolate to tetrahydrofolate. The latter substance is important to the biosynthesis of purines, pyrimidines and several of the amino acids and is therefore necessary for normal cell function (Blakley, 1984). DHFR is a ubiquitous protein, found not only in bacteria but also man, and one of the unique properties of TMP is its ability to inhibit selectively the bacterial species of the enzyme (Roth, 1983; Hitchings et. al., 1988). That remarkable selectivity is illustrated in Table 1.

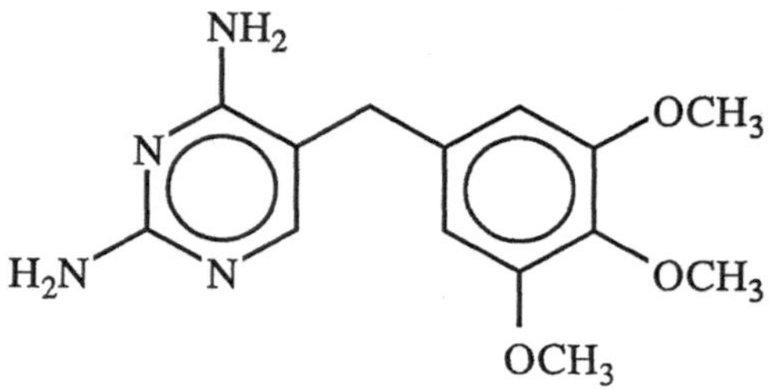

1

57

Table 1. Trimethoprim inhibition data for a selected set
of bacterial and vertebrate DHFR

Enzyme source	I_{50} (nM)[a]	K_i (nM)[b]
Escherichia coli	8.	0.08
Staphylococcus aureus	5.	
Proteus vulgaris	7.	
Mouse (SR1)	280,000.	
Chicken	470,000.	
Human	490,000.	960.

[a]Data from R. Ferone and D. Baccanari, Wellcome Research
Laboratories.

[b]Appleman et al., 1988.

TMP exhibits potent inhibition of most bacterial forms of
DHFR, but its effect on the vertebrate species is
dramatically weaker. The ratio of K_i values for the *E. coli*
and human enzymes is 12,000 (Appleman et al., 1988) and
corresponds to a difference in binding free energy of
5.6 kcal/mol. That energy difference represents one of the
key factors that makes TMP a useful antibacterial agent,
and, based on inhibition data from several hundred
analogues (see Blaney et al., 1984), appears to be
intimately associated with the aromatic methoxy groups of
the drug, as exemplified by the data shown in Table 2.

Methoxy substitution has a striking effect on both
antibacterial activity and DHFR selectivity. As seen in
Table 2, the unsubstituted benzylpyrimidine is a relatively
weak inhibitor of *E. coli* DHFR, but activity against that
enzyme is significantly higher for the methoxy-substituted
derivatives. The two monomethoxy analogues are each about
10-fold more active than the parent inhibitor, the two
dimethoxy compounds are about 100-fold more active, and TMP
is almost three orders of magnitude more active. In
contrast, the mouse enzyme (as well as DHFR from other
vertebrate sources including man) is relatively insensitive
to methoxy substitution: all of the compounds listed in
Table 2 show approximately the same weak inhibition of

vertebrate DHFR. Thus the observed selectivity parallels the activity against the bacterial enzyme, so that TMP is not only the most potent inhibitor of *E. coli* DHFR in this series of inhibitors but is also the most selective. Clearly, the aromatic methoxy group plays an important role in DHFR selectivity for this class of inhibitor, and the following discussion relates some of our efforts to understand the basis for that effect. Perhaps such understanding will provide an advantage in the design of new types of selective inhibitor.

Because of its therapeutic importance, DHFR has received much attention during the last three decades and has been studied extensively by x-ray crystallography (for reviews see Kuyper, 1989; Kraut et al., 1987; Beddell, 1984; Freisheim et al., 1984;). Three-dimensional structures of five different DHFR complexes with TMP are now known, three of which have been refined at reasonably high resolution. Those structures, listed in Table 3, offer the opportunity to gain insight into the binding and selectivity of TMP,

Table 2. DHFR inhibition data for a series of TMP analogues[a]

| | DHFR I$_{50}$ (nM) | | |
R	*E. coli*	Mouse (SR1)	Selectivity[b]
H	4100.	180,000.	44.
3-OMe	450.	190,000.	420.
4-OMe	480.	160,000.	330.
3,4-diOMe	51.	73,000.	1,400.
3,5-diOMe	53.	220,000.	4,200.
3,4,5-triOMe (TMP)	8.	280,000.	35,000.

[a]Baccanari et al., 1982.
[b]Mouse DHFR I$_{50}$ / *E. coli* DHFR I$_{50}$

Table 3.   Reported x-ray crystal structures of DHFR in
           complex with TMP.

Enzyme source	Ligands	Resolution (Å)	Reference
E. coli	TMP	2.3 refined	Matthews et al., 1985a
E. coli	TMP, NADPH	3.0	Champness et al., 1986b
Chicken	TMP, NADPH	2.2 refined	Matthews et al., 1985b
Mouse	TMP, NADPH	2.0 refined	Stammers et al., 1987
Human	TMP	3.5	Oefner et al., 1988

and several papers have appeared on that subject. In
particular, Matthews et al. (1985a and b) have published an
elegant and thorough comparison of the *E. coli* DHFR-TMP
binary complex and the corresponding ternary (DHFR-TMP-
NADPH) complex of the chicken enzyme.   Those same two
structures, as well as hypothetical models of the
respective ternary and binary complexes, were studied by
Roberts et al. (1986) using molecular mechanics energy
minimization and interaction energy analysis. The analysis
presented below builds on that published work by including
the x-ray structures of the *E. coli* (Champness et al., 1986b)
and mouse DHFR (Stammers et al., 1987) ternary complexes.
The ternary complex is believed to be the biologically
relevant form of the inhibited enzyme, and cooperative
binding of NADPH has been shown to be important to the
activity and selectivity of TMP (Baccanari et al., 1982).
In addition, the conformation of a protein loop that is
near the nicotinamide portion of the cofactor is different
in the binary and ternary *E. coli* DHFR crystal structures and
alters the overall interaction between protein and TMP in
the two complexes, an observation that was anticipated by
Matthews et al. (1985a).   We have therefore focused our
analysis on the three ternary complexes listed in Table 3.
In the energy analysis of Roberts et al. (1986), a model
for the ternary *E. coli* DHFR complex was constructed using
crystal structure coordinates of the binary complex, assuming

incorrectly that NADPH binding would have no effect on protein conformation. The calculated interaction energy between TMP and protein in that model was therefore misleading. Our analysis also includes an additional consideration of solvation, and, as discussed below, it appears that solvation may play an important role in the DHFR selectivity of TMP.

The amino acid sequences of DHFR from vertebrate sources are highly homologous (75-90%), and the three-dimensional structures of the chicken, mouse and human enzymes show a strong similarity to one another (see Champness, et al., 1986a). In contrast, sequence similarity between bacterial and vertebrate species of the enzyme is low, in the range of 20-30%. Nonetheless, these two species of DHFR show three-dimensional properties that have many features in common.

Each DHFR structure is dominated by an 8-stranded beta-sheet, as illustrated for the *L. casei* enzyme in Figure 1. The sheet is flanked on either side by two alpha helices, and the active site is found between helices B and C. Vertebrate DHFR is generally about 25-30 residues larger than bacterial protein, but the majority of those extra residues are accommodated as loops at the surface of the enzyme and appear to have little effect on overall folding. The active site regions of the two protein species are relatively similar in spite of these differences in sequence length and composition.

However, the bacterial and vertebrate DHFR binding cavities do differ in subtle ways, as described by Matthews et al. (1985b), and it is those subtle differences in size and shape of the binding sites that give rise to the large difference in affinity for TMP and also a significant difference in binding conformation for the drug, as shown in Figures 2 and 3. The diaminopyrimidine ring of TMP associates with each species of DHFR in about the same manner, binding deep inside the enzyme cleft through several hydrogen bonds. The obvious difference in the two TMP complexes is the conformation of inhibitor. In the bacterial enzyme complex, the dihedral angle about the pyrimidine-to-methylene bond is about 180 degrees, but in the corresponding vertebrate structure that angle is about 270 degrees. The trimethoxyphenyl group is therefore positioned in different regions of the two active sites.

Matthews et al. (1985b) have rationalized this difference in TMP binding conformation in terms of a size differential for the two types of DHFR binding cleft. To summarize their reasoning very briefly, they observed that the trimethoxyphenyl group of TMP appeared to be favorably accommodated in the lower part of the *E. coli* DHFR binding cavity, tightly sandwiched between a number of hydrophobic residues, and the drug's diaminopyrimidine ring was able to form a full complement of hydrogen bonds to the protein deep inside the cleft. In the vertebrate enzyme the binding cleft is somewhat larger, and the trimethoxyphenyl group of the drug is unable to interact favorably with both sides of the lower part of the cleft. The larger vertebrate DHFR binding cavity provides a more favorable

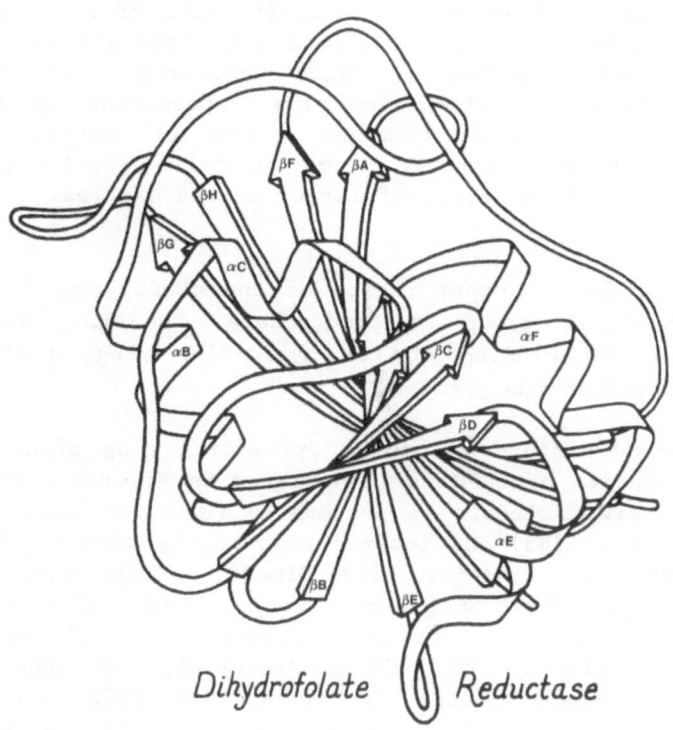

Figure 1. Ribbon representation of *L. casei* DHFR, reproduced with permission from Jane Richardson and Academic Press (Richardson, 1981).

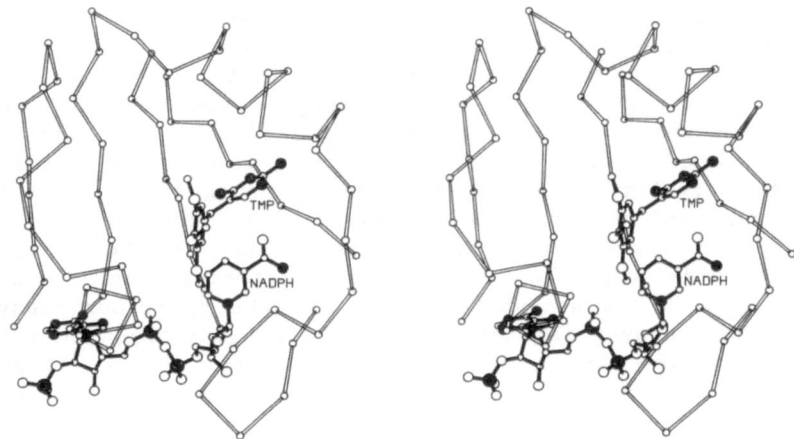

Figure 2. Stereo diagram of the active site of *E. coli* DHFR–
NADPH–TMP. Protein α–carbon structure is shown
with open bonds, ligand structures with filled
bonds. Carbon atoms are represented by small
open circles, oxygen atoms by larger open
circles, nitrogen atoms by filled circles, and
phosphorus atoms of larger filled circles.

environment in the upper region of the cleft, but for the
trimethoxyphenyl group to bind there, the drug has to
compromise the binding of its pyrimidine ring, losing a
hydrogen bond in the process.

The observed difference in TMP pyrimidine ring binding in
the two species of DHFR is illustrated in Figures 4 and 5.
The bound drug molecule is protonated (Bevan et al., 1985;
Cocco et al., 1983) and interacts ionically with Asp-27 of
E. coli DHFR and likewise with Glu-30 of the vertebrate
enzymes. The 2-amino group hydrogen bonds to a buried
water molecule that has been observed in all refined
crystal structures of DHFR. Presumably an analogous
solvent molecule exists in the *E. coli* DHFR–NADPH–TMP
complex: that structure has not been refined but almost
certainly accommodates that water molecule, based on
similarities of protein geometry. In the *E. coli* enzyme, the

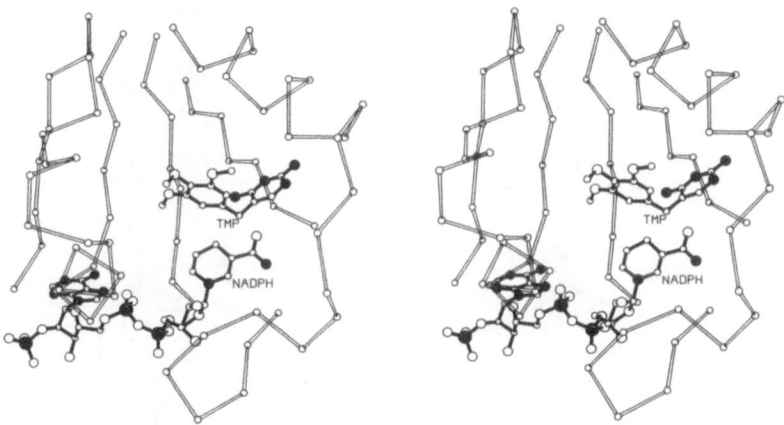

Figure 3. Stereo diagram of the active site of chicken
DHFR-NADPH-TMP. See caption for Figure 2.

4-amino group donates hydrogen bonds to two backbone
carbonyl units, whereas in the vertebrate protein complex
only one hydrogen bond is observed because of the somewhat
different position of the pyrimidine ring. The distance
between the amino group nitrogen and the carbonyl oxygen of
Val-115 in vertebrate DHFR is 4.1-4.5 Å, too long for an
ideal hydrogen bond. Non-selective inhibitors such as
methotrexate bind to vertebrate DHFR with their pyrimidine
ring positioned in a manner analogous to that of TMP bound
to bacterial DHFR, that is with the full complement of five
hydrogen bonds (Stammers et al., 1987; Matthews et al.,
1985b). Only TMP and two of its close analogues have shown
the altered pyrimidine binding.

An important question is how much does the observed
difference in pyrimidine binding contribute to the
5.6 kcal/mol of differential binding free energy of TMP.
This issue has been addressed by Matthews et al. (1985) and
also Roberts et al. (1987). It's clearly difficult to
evaluate quantitatively that contribution to selectivity,
but whatever that may be, it's also difficult to
rationalize the effect of methoxy substitution illustrated
in Table 2 in terms of pyrimidine binding alone. We

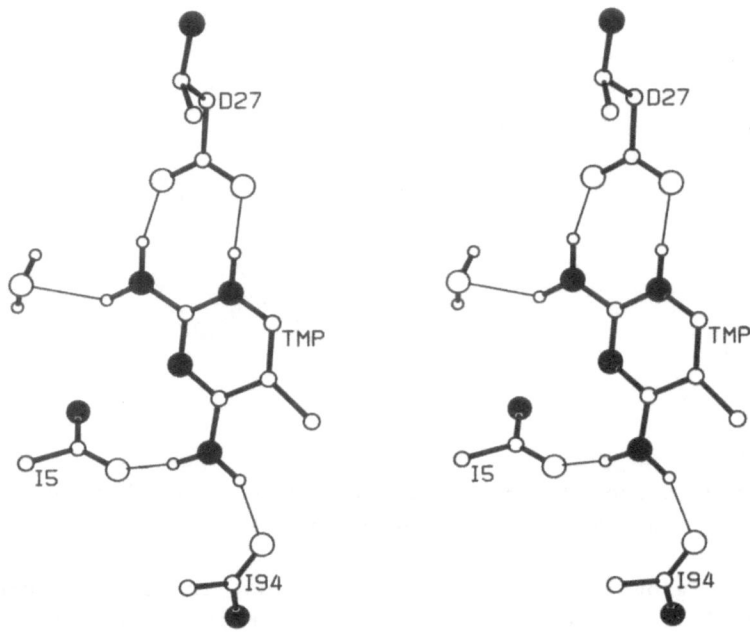

Figure 4. Stereo diagram of the hydrogen bond interactions of the diamino-pyrimidine ring of TMP in its complex with *E. coli* DHFR. Hydrogen atoms are represented by the small open circles, carbon atoms by the larger open circles, oxygen atoms by the largest open circles, and nitrogen atoms by the filled circles.

therefore attempted to analyze the interactions between the trimethoxybenzyl group and the DHFR-NADPH binding site in both species of enzyme.

As shown in Figure 6, the trimethoxybenzyl group of TMP is found in a hydrophobic region of the *E. coli* DHFR active site, surrounded by sidechains from Met-20, Leu-28, Phe-31, Ile-50, and Leu-54. Similar non-specific interactions are observed in the vertebrate enzyme complex, as shown in Figure 7. Simple visual inspection of the two complexes gave little insight into why methoxy groups are so

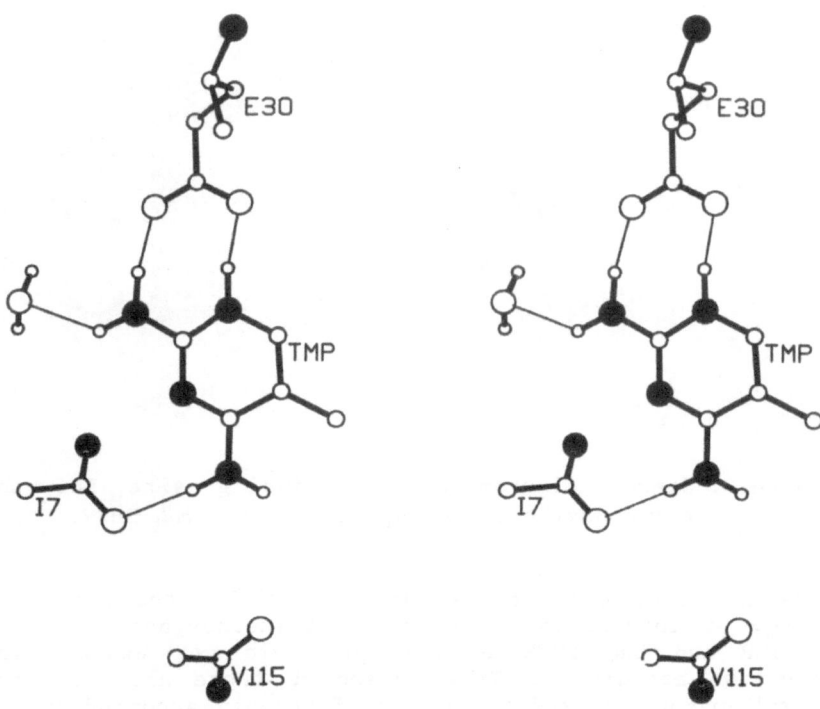

Figure 5. Stereo diagram of the hydrogen bond interactions
of the diamino-pyrimidine ring of TMP in its
complex with chicken DHFR. See caption of Figure 5.

important to selectivity. A more quantitative analysis
using molecular mechanics interaction energies also (see
Pettitt et al., 1986) failed to rationalize the methoxy
group effect. The interaction energy between the
trimethoxybenzyl moiety of TMP and the DHFR-NADPH binding
site was calculated for each enzyme complex using atomic
coordinates of the x-ray structure as well as coordinates
from energy minimized versions of those complexes. Energy
minimization was carried out with harmonic restraints
applied to each atom so that geometry deviations from the
experimental structure were small (see Bruccoleri et al.,
1986). The overall RMS deviations of the energy minimized
structures were less than 0.17 Å, a value well within the
estimated experimental error of the x-ray structures.
Regardless of which versions of the structures

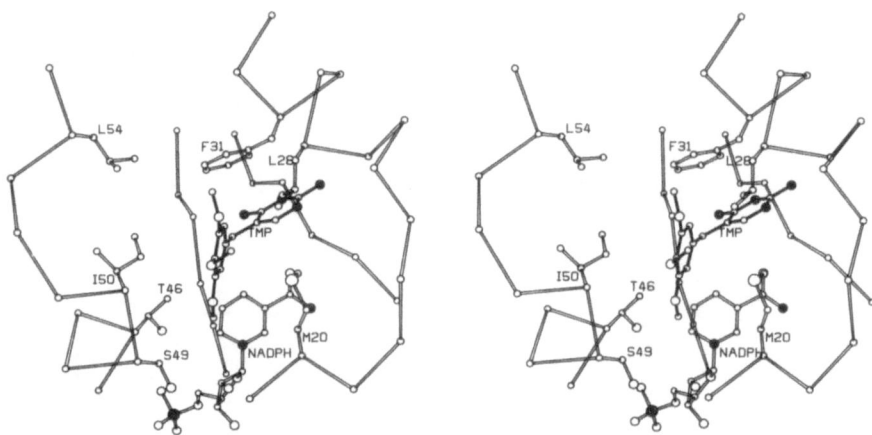

Figure 6. Stereo diagram of the binding site for the trimethoxybenzyl group of TMP in *E. coli* DHFR.

were analyzed, the results were essentially the same: the energy of interaction between the trimethoxybenzyl moiety of TMP and the DHFR-NADPH complex did not explain the selective affinity of TMP. Based on these analyses, the benzyl group appeared to be more favorably accommodated in vertebrate enzyme than in bacterial DHFR. The contribution of van der Waals interactions was similar in the two types of complex, but electrostatic energies suggested that the vertebrate DHFR environment was somewhat better for binding the drug's benzyl group. Of course, electrostatic interactions are often difficult to evaluate because of complications associated with dielectric constant and with charged residues at the protein surface that should be solvated and probably paired with counterions. However, no matter how we addressed those problems, the coulombic interaction terms always favored the vertebrate enzyme environment. Our conclusion was that direct interactions between the trimethoxybenzyl group and DHFR-NADPH did not appear to be responsible for species selectivity of TMP.

As discussed above, the conformation of TMP bound to bacterial DHFR is different from that of the vertebrate DHFR complex, and there was a possibility that conformational strain energy might play a role in selectivity. That possibility has been addressed by both

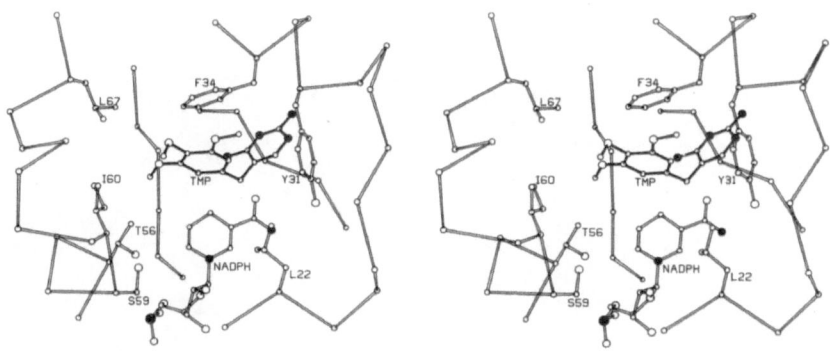

Figure 7. Stereo diagram of the binding site for the
trimethoxybenzyl group of TMP in chicken DHFR.

Matthews et al. (1985b) and Roberts et al. (1986), and both
groups concluded that the two conformations were similar in
stability and probably did not contribute to selective
binding. Our analysis agrees with that view and is
discussed only as a prelude to the additional
conformational analyses that follow.

Results from full relaxation molecular mechanics analysis
of TMP are depicted in Figure 8. The molecule is
relatively flexible through rotation about the two central
single bonds, and the corresponding energy contour map
shows an extended energy valley with four minima. As seen
in Figure 8, the conformations observed for TMP in the two
vertebrate enzyme complexes are close to one of the
calculated minima, and those for the bacterial enzyme
complexes are about 1 kcal/mol less stable. Within the
accuracy of these calculations, the two binding
conformations are of similar stability, and if there is a
difference it appears to favor the vertebrate enzyme
complex.

We have also analyzed the conformational preferences of TMP
and the set of desmethoxy analogues listed in Table 2 in
the enzyme-bound state. As shown in Figure 9, the
calculated minimum energy conformation of TMP bound to
E. coli DHFR is similar to the observed conformation,
providing some confidence in our method of calculation.
The two minima in Figure 9 correspond to alternative

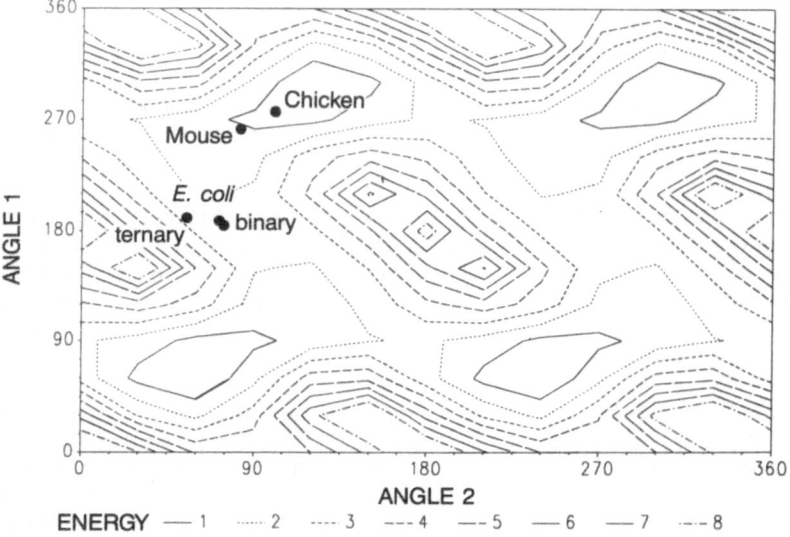

Figure 8. Molecular mechanics energy contour map for TMP
with conformations of the enzyme bound inhibitor
indicated.

conformations of the para-methoxy group, the methyl moiety
of which is forced out of the phenyl plane by the adjacent
substituents. When applied to the desmethoxy analogues,
this modeling procedure predicted that those inhibitors
bind to the bacterial enzyme in essentially the same manner
as observed for TMP, as illustrated for the parent
benzylpyrimidine in Figure 10.

Modeling was also carried out on the mouse DHFR-NADPH
complexes of TMP and its desmethoxy analogues. The results

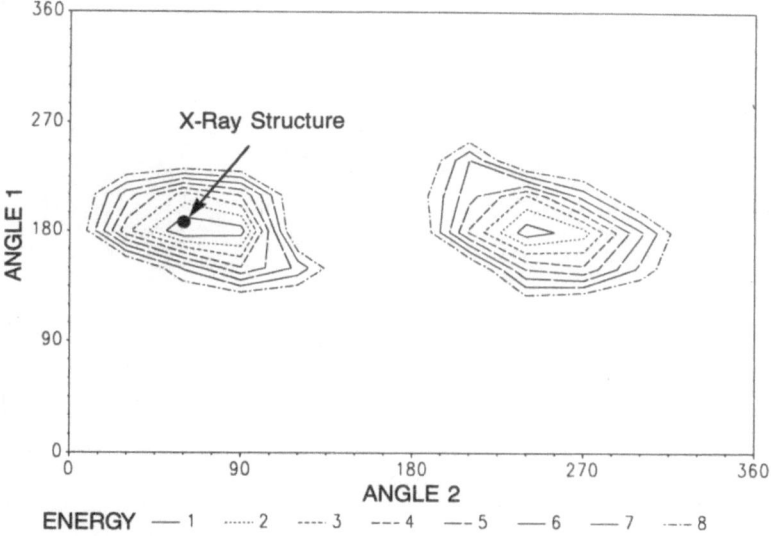

Figure 9. Molecular mechanics energy contour map calculated for TMP in complex with *E. coli* DHFR-NADPH with the conformation observed in the corresponding x-ray structure indicated.

for TMP are shown in Figure 11, and the calculated minimum was again reasonably close to the observed conformation. As for the *E. coli* DHFR complexes, each of the TMP analogues predicted to bind to mouse DHFR-NADPH in a mode similar to that found for TMP, including the position of the diaminopyrimidine ring. The corresponding energy contour map for the unsubstituted analogue is shown in Figure 12.

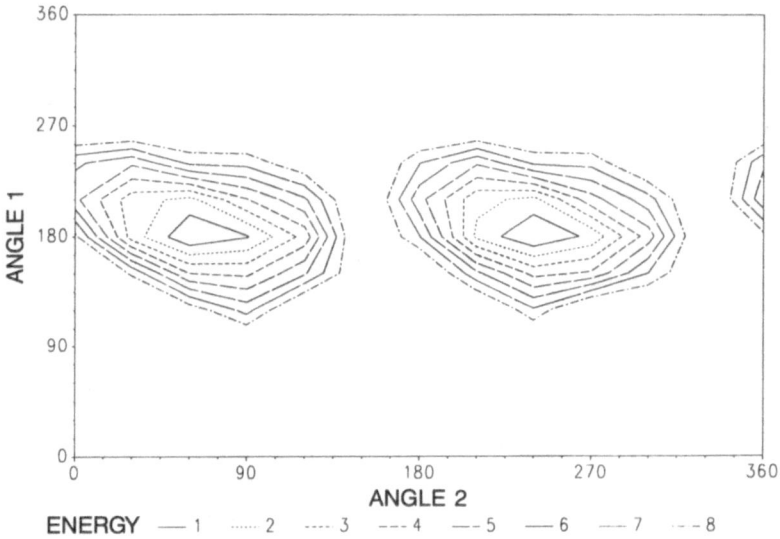

Figure 10. Molecular mechanics energy contour map calculated for 2,4-diamino-5-benzylpyrimidine in complex with *E. coli* DHFR-NADPH.

Interaction energy analysis of the hypothetical enzyme complexes with the desmethoxy analogues of TMP gave essentially the same results that we obtained for TMP. The energy of interaction between the benzyl group of each inhibitor and DHFR-NADPH was similar for the bacterial and vertebrate species and did not provide an explanation for the effect of methoxy substitution on selectivity. These

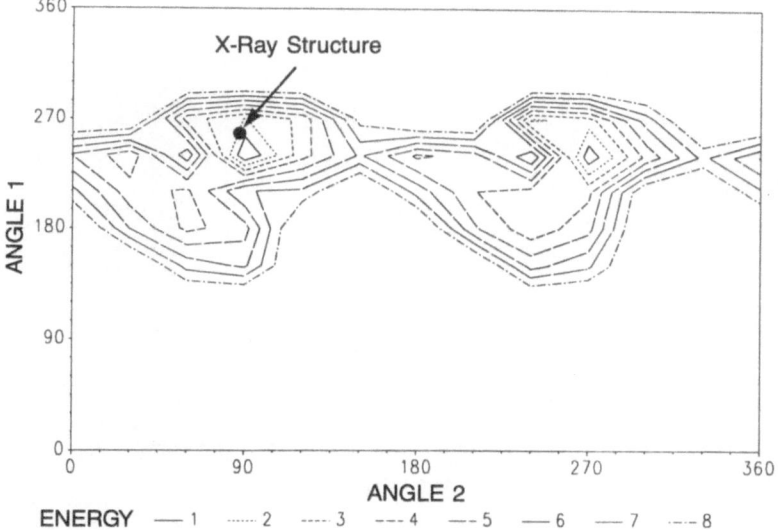

Figure 11. Energy contour map for TMP bound to mouse DHFR-
NADPH with the conformation observed in the
corresponding x-ray structure indicated.

results suggested that the methoxy group effect was due to
some aspect of enzyme binding that was not addressed in our
analysis up to this point. One obvious omission is the
effect of solvent.

In considering the potential effects of solvation on TMP
binding to DHFR, the solvation energy associated with an
aromatic methoxy group is of interest. To our knowledge
the only literature data available are found in a

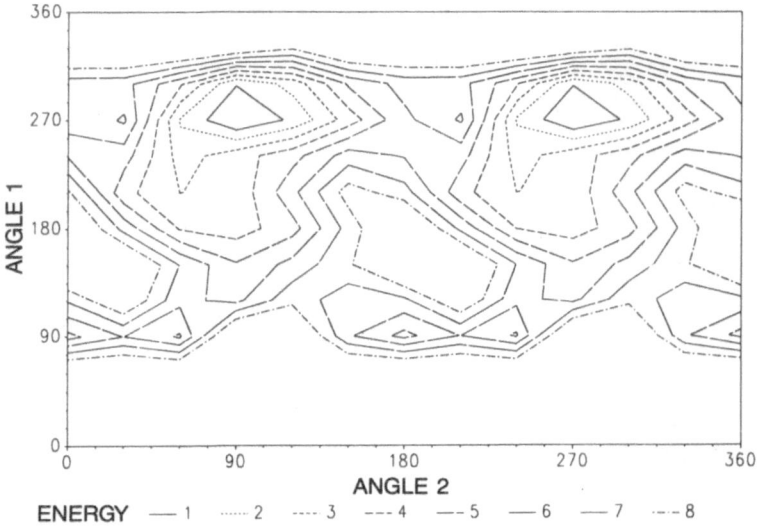

Figure 12. Energy contour map for 2,4-diamino-5-benzyl-pyrimidine bound to mouse DHFR-NADPH.

compilation of solvation data reported by Hine et al. (1975). Based on the reported gas phase/water partitions of anisole and benzene, the free energy of solvation for the aromatic methoxy group was only -0.1 kcal/mol. However, we re-determined the solvation energy of anisole and also measured solvation energies for 1,2-dimethoxybenzene and 1,2,3-trimethoxybenzene, as shown in Table 4, and those data indicated a significant methoxy group effect on

Table 4. Water solubilities, vapor pressures and solvation energies of benzene and several methoxy-substituted derivatives.

Compound	Water Solubility(M)	Vapor Pressure(mm Hg)	Solvation Energy(kcal/mol)
benzene	0.023	95.	−0.89
anisole (—OCH$_3$)	0.013	4.2	−2.4
dimethoxybenzene (OCH$_3$, —OCH$_3$)	0.046	1.5	−3.8
trimethoxybenzene (OCH$_3$, —OCH$_3$, OCH$_3$)	0.030	0.059	−5.4

solvation energy.   Solvation energies were determined in standard fashion (Hine et al., 1975) by measuring water solubilities and vapor pressures to give the gas phase/water partition values. Our data showed that methoxy group contribution to solvation free energy was about −1.5 kcal/mol, the negative value indicating a preference for water versus the gas phase. The solvation energy of trimethoxybenzene is −5.4 kcal/mol, an energy value that is close to the 5.6 kcal/mol selectivity difference in the DHFR binding free energy of TMP.   The magnitude of that energy suggests that solvation could make a significant contribution to selectivity.

Solvation may contribute to TMP selectivity in terms of differential desolvation of the inhibitor.   In the bacterial enzyme complex, the oxygen atoms of the methoxy groups of the drug are at the cleft opening and are relatively accessible to solvent.   Solvent accessible surface area calculations show that the oxygen atom of each methoxy group of TMP bound to *E. coli* DHFR-NADPH can directly contact solvent, as represented by a 1.4 Å sphere. In contrast, the oxygen atoms of TMP have no solvent accessibility in the drug's complex with either mouse or chicken DHFR-NADPH, suggesting that the inhibitor must undergo desolvation to a significantly greater extent to bind to the vertebrate enzyme than to bacterial protein.

To further explore the differential interaction between TMP and water in the two types of DHFR complexes, we constructed a solvated model of each complex in which a spherical cap of water molecules (16 Å radius, centered on the para-methoxy oxygen of TMP) was placed over the active site. The water cap was then subjected to 6 psec of molecular dynamics followed by energy minimization, and the energy of interaction between the water cap and the trimethoxybenzyl group in each complex was calculated. This procedure was performed twice in each case, and the results were consistent. The average difference in the interaction energy between the bacterial and vertebrate enzyme complexes was 17 kcal/mol, with the bacterial DHFR-bound version of TMP showing the more favorable interaction. A simple interaction energy cannot be compared directly to a free energy, but the magnitude of differential interaction energy is consistent with a significant difference in the association between TMP and water in the two types of enzyme complexes. To obtain a more realistic value for this interaction energy, the energy was monitored during a molecular dynamics simulation. The average difference in interaction energy during 10 psec simulations at 300°K of the two types of complex was 4 kcal/mol, a value more in line with expectation.

To summarize and possibly clarify the potential role of solvation in the effect of methoxy group substitution on DHFR selectivity, we offer the following rationalization of the inhibition data shown in Table 2. The parent unsubstituted benzylpyrimidine is relatively non-selective, and although reliable vertebrate Ki values for this compound are not available, its selectivity is estimated to correspond to 1-2 kcal/mol of binding free energy. If the molecular modeling calculations are accurate and the mode of binding for this parent compound and each methoxy-substituted derivative is essentially the same as observed for TMP, then perhaps the DHFR selectivity of the parent inhibitor arises primarily from the differential binding of the pyrimidine ring. That basal selectivity is then augmented by methoxy substitution. The observed increase of about 10-fold (1.4 kcal/mol) in affinity for the *E. coli* enzyme per methoxy group presumably results from favorable interactions with protein and the solvent-exposed nature of the methoxy substituents in the complexes with that enzyme. The energy cost in desolvation for binding to the bacterial enzyme is relatively small. Conversely, in binding to vertebrate DHFR each methoxy group also appears to be favorably accommodated by the protein binding cavity, perhaps involving a contribution to binding free energy similar to that found for the bacterial enzyme complexes— about 1.4 kcal/mol. However, those interactions with vertebrate protein take place at the expense of methoxy group desolvation, a process that may cost about 1.5 kcal/mol. The protein binding and desolvation effects on inhibitor affinity cancel one another, and the overall result is that methoxy substitution has little effect on binding to vertebrate DHFR. Thus differential desolvation would appear to be an important contributor to the selective DHFR affinity of TMP.

The above rationalization is perhaps an oversimplication; however, it does identify the potential significance of solvation to inhibitor selectivity and further illustrates the need for careful consideration of solvent effects in the design of new inhibitors.

Acknowledgements

The author is grateful to three groups for x-ray crystallographic data on DHFR-TMP complexes: D. Matthews, J. Kraut and coworkers at the University of California, San Diego; C. Beddell, J. Champness, D. Stammers and coworkers at the Wellcome Research Laboratories, Beckenham, England; and A. Geddes, A. North and coworkers at the University of Leeds, Leeds, England. Molecular mechanics and dynamics calculations were performed with AMBER software that was developed by the group of Peter Kollman at the University of California, San Francisco (AMBER Version 3.0, see Weiner et. al., 1981). We thank Professor Kollman for help and advice with those calculations. We also made use of the program MacroModel, which comes from the group of Clark Still at Columbia University. In the determination of solvation energies, vapor pressures were measured by David Ashton of Wellcome Research Laboratories,

76

Beckenham, England, and water solubilities were determined by Jane Muse, Doug Minick and Robert Hunter of Burroughs Wellcome Co.

References

Appleman, J.R., Prendergast, N., Delcamp, T.J., Fresheim, J.H., and Blakley, R.L., 1988, Kinetics of the formation and isomerization of methotrexate complexes of recombinant human dihydrofolate reductase, J. Biol. Chem., 263:10304-10313.

Baccanari, D.P., Daluge, S., and King, R.W., 1982, Inhibition of dihydrofolate reductase: Effect of reduced nicotinamide adenine dinucleotide phosphate on the selectivity and affinity of diaminobenzylpyrimidines, Biochem., 21:5068-5075.

Beddell, C.R., 1984, Dihydrofolate reductase: Its structure, function, and binding properties, in "X-ray Crystallography and Drug Action", (A.S. Horn and C.J. DeRanter, eds.), pp. 169-193, Oxford University Press, New York.

Bevan, A.W., Roberts, G.C.K., Feeney, J. and Kuyper, L.F., 1985, 1H and 15N NMR studies of protonation and hydrogen-bonding in the binding of trimethoprim to dihydrofolate reductase, Eur. Biophys. J., 11:211-218.

Blakley, R.L., 1984, Dihydrofolate Reductase, in "Chemistry and Biochemistry of Folates" (R. L. Blakley and S. J. Benkovic, eds.), pp. 191-253, John Wiley & Sons, New York.

Blaney, J.M., Hansch, C., Silipo, C., and Vittoria, A., 1984, Structure- activity relationships of dihydrofolate reductase inhibitors, Chem. Rev., 84:333-407.

Bruccoleri, R.E. and Karplus, M., 1986, Spatially constrained minimization of macromolecules, J. Comp. Chem., 7:165-175.

Champness, J.N., Kuyper, L.F., and Beddell, C.R., 1986a, Interaction between dihydrofolate reductase and certain inhibitors, in "Molecular Graphics and Drug Design" (A.S.V. Burgen, G.C.K. Roberts, and M.S. Tute, eds.), pp. 335-362, Elsevier, Amsterdam.

Champness, J.N., Stammers, D.K. and Beddell, C.R., 1986b, Crystallographic investigation of the cooperative interaction between trimethoprim, reduced cofactor and dihydrofolate reductase, FEBS Lett., 199:61-67.

Cocco, L., Roth, B., Temple, C., Jr., Montgomery, J.A., London, R.E., and Blakley, R.L., 1983, Protonated state of methotrexate, trimethoprim, and pyrimethamine bound to dihydrofolate reductase, Arch. Biochem. Biophys., 226:567-577.

Finland, M., Kass, E.H., and R. Platt (eds.), 1982, Trimethoprim-sulfamethoxazole revisited, Reviews of Infectious Diseases, 4:196-618.

Freisheim, J.H. and Matthews, D.A., 1984, The comparative biochemistry of dihydrofolate reductase, in "Folate Antagonists as Therapeutic Agents", Vol. 1, (F.M. Sirotnak, J.J. Burchall, W.B. Ensminger, and J.A. Montgomery, eds.), pp. 69-131, Academic Press, Orlando.

Hine, J. and Mookerjee, P.K., 1975, The intrinsic hydrophilic character of organic compounds. Corre ations in terms of structural contributions, J. Org. Chem., 40: 292-298.

Hitchings, G.H., 1983, Functions of tetrahydrofolate and the role of dihydrofolate reductase in cellular metabolism, in "Inhibition of Folate Metabolism in Chemotherapy: The Origins and Uses of Co-trimoxazole", (G.H. Hitchings, ed.), pp. 11-23, Springer-Verlag, Berlin.

Hitchings, G.H., Kuyper, L.F., and Baccanari, D.P., 1988, Selective inhibitors of dihydrofolate reductase, in "Design of Enzyme Inhibitors as Drugs", (M. Sandler and H.J. Smith, eds.), pp. 343-362, Oxford University Press, Oxford.
Kraut, J. and Matthews, D.A., 1987, Dihydrofolate reductase, in "Active sites of Enzymes", (F. Jurnak and A. McPherson, eds.), pp. 1-71, Wiley and Sons, New York.

Kuyper, L.F., 1989, Inhibitors of dihydrofolate reductase, in "Computer-Aided Drug Design", (T.J. Perun and C.L. Propst, eds.), pp. 327-369, Marcel Dekker, Inc., New York.

Matthews, D.A., Bolin, J.T., Burridge, J.M., Filman, D.J., Volz, K.W., Kaufman, B.T., Beddell, C.R., Champness, J.N., Stammers, D.K., and Kraut, J., 1985a, Refined crystal structures of Escherichia coli and chicken liver dihydrofolate reductase containing bound trimethoprim, J. Biol. Chem., 260:381-391.

Matthews, D.A., Bolin, J.T., Burridge, J.M., Filman, D.J., Volz, K.W. and Kraut, J., 1985b, Dihydrofolate reductase. The stereochemistry of inhibitor selectivity, J. Biol. Chem., 260:392-399.

Oefner, C., D'Arcy, A., and Winkler, F.K., 1988, Crystal structure of human dihydrofolate reductase complexed with folate, Eur. J. Biochem., 174:377-385.

Pettitt, M. and Karplus, M., 1986, Interaction energies: their role in drug design, in "Molecules Graphics and Drug Design", (A.S.V. Burgen, G.C.K. Roberts, and M.S. Tute, eds.), pp. 75-113, Elsevier, New York.

Richardson, J.S., 1981, The anatomy and taxonomy of protein structure, Adv. Protein Chem., 34:167-339.

Roberts, V.A., Dauber-Osguthorpe, P., Osguthorpe, D.J., Levin, L., and Hagler, A.T., 1986, A comparison of the binding of the ligand trimethoprim to bacterial and vertebrate dihydrofolate reductases, Israel J. Chem., 27:198-210.

Roth, B., 1983, Selective inhibitors of bacterial dihydrofolate reductase: Structure-activity relationships, in "Inhibition of Folate Metabolism in Chemotherapy: The Origins and uses of Co-trimoxazole", (G.H. Hitchings, ed.), pp. 107-127, Springer-Verlag, Berlin.

Stammers, D.K., Champness, J.N., Beddell, C.R., Dann, J.G., Eliopoulos, E., Geddes, A.J., Ogg, D., and North, A.C.T., 1987, The structure of mouse L1210 dihydrofolate reductase-drug complexes and the construction of a model of human enzyme, FEBS Lett., 218:178-184.

Weiner, P.K. and Kollman, P.A., 1981, AMBER: Assisted model building with energy refinement. A general program for modeling molecules and their interactions, J. Comp. Chem., 2:287-303.

Crystallographic and Genetic Approaches Toward the Design of Proteins of Enhanced Thermostability

J.A. WOZNIAK, X.-J. ZHANG, K. WILSON,
L.H. WEAVER, D.E. TRONRUD, P.E. PJURA,
H. NICHOLSON, M. MATSUMURA, M. KARPUSAS,
R. JACOBSON, R. FABER, S. DAO-PIN, J.A. BELL,
T. ALBER, and BRIAN W. MATTHEWS

INTRODUCTION

The advent of directed mutagenesis has made it possible to alter protein structures at will. For the first time it is possible to design and to introduce modifications into a protein that are intended to change its behavior in predictable ways.

We have been using the lysozyme from bacteriophage T4 as a model system to test ways in which the stability of a protein might be improved. Such studies also provide quantitative information on the contributions that different types of interactions (H–bonds, hydrophobic interactions, salt bridges, etc.) make to the stability of proteins (Matthews, 1987). As such, these studies are also relevant to the contributions that these different interactions can make in enzyme-inhibitor and drug–receptor complexes.

In this chapter we will briefly review some of the approaches that are being explored in an attempt to increase the thermostability of T4 lysozyme. A very similar report has been given by Bell et al. (1989).

This chapter has been reprinted, with permission, from *The Use of X-Ray Crystallography in the Design of Anti-Viral Agents*, edited by G.M. Air and W.G. Laver, published by Academic Press, January 1990.

Hydrophobic interactions

It has long been known that the hydrophobic effect plays a very important role in stabilizing protein structures (Kauzmann, 1959).

In order to quantitate the contribution of the hydrophobic effect at a specific site in a protein, isoleucine 3 in T4 lysozyme was replaced with 13 different amino acids (Matsumura, Becktel and Matthews, 1988). It was found that the contributions of different residues at position 3 to the overall stability of the protein were directly proportional to the hydrophobicity of the substituted residue (Figure 1). Phenylalanine, tyrosine and tryptophan were exceptions because, as inferred from the crystal structure of the Tyr variant, the side-chains of these residues are too large to be accommodated within the interior of the protein.

It was of interest to find that two substitutions at position 3 yielded mutant proteins with thermostability greater than that of wild-type lysozyme. The first thermostable variant has cysteine at position 3, and the enhanced stability can be attributed to the formation of a non-native disulfide bridge (Perry and Wetzel, 1984) (see below). The second variant more stable than wild-type has leucine at position 3, and the enhanced stability is presumed to be due to increased hydrophobic stabilization. Because crystals could not be obtained the structure of the Leu 3 variant has not been determined. Model building suggests that the Leu 3 side-chain is completely buried within the protein, whereas the side-chain of Ile 3 in wild-type lysozyme is 15% exposed to solvent and therefore does not manifest its full hydrophobic potential (Matsumura, Becktel and Matthews, 1988; Matsumura et al., 1989b). This result suggests that there may be other sites within the T4 lysozyme molecule (or within proteins in general) where enhanced hydrophobic stabilization

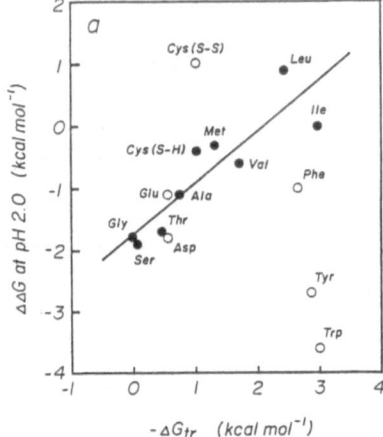

Figure 1. Stabilities of lysozymes with 13 different amino acid substitutions at Ile 3. The free energy of stabilization ($\Delta\Delta G$) of each mutant lysozyme at pH 2.0 is plotted against the free energy of transfer ($-\Delta G_{tr}$) of the individual amino acid from water to ethanol. The protein free energies are plotted relative to wild-type (Ile). (Reprinted by permission from <u>Nature</u> <u>334</u>, 406-410. Copyright (C) 1988 Macmillan Magazines Ltd.)

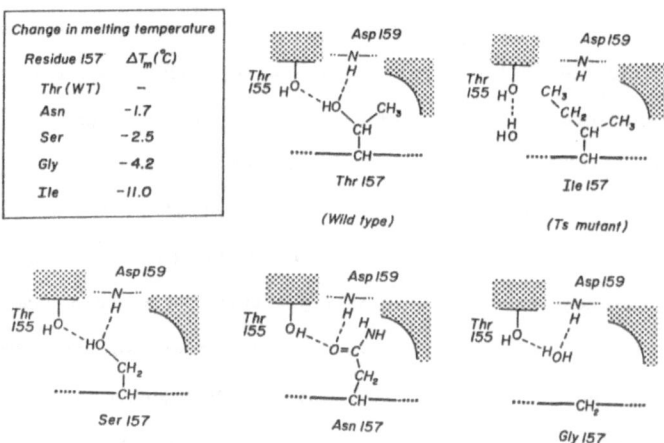

Figure 2. Schematic illustration showing the interactions displayed by five representative amino acids at position 157 of T4 lysozyme. The insert shows the relative melting temperatures at pH 2.0 of the mutant lysozymes illustrated in the figure. (Adapted from Alber et al., 1987.)

might be achieved by appropriately chosen amino acid replacements. In an attempt to find such sites a systematic search was made for cavities within the T4 lysozyme structure that might permit the replacement of smaller hydrophobic amino acids by larger ones, and so achieve greater hydrophobic stabilization (Karpusas et al., manuscript in preparation). There are such cavities, and one of these appears to allow the replacement of Leu 133 by a phenylalanine. This mutant protein has been constructed and high resolution crystallographic analysis confirms that the Phe 133 side-chain does occupy the cavity as expected. The crystallographic analysis also shows, however, that the side-chain rotational angle, χ^1, is not optimal, and the strain energy associated with this distortion appears to offset the hydrophobic stabilization expected from the Leu → Phe replacement. As a result, the measured thermostability of the mutant protein is essentially identical with wild-type.

Hydrogen bonding

The importance of hydrogen bonding in determining the folded structures of proteins does not need to be emphasized. On the other hand, the energetic contributions of individual hydrogen bonds are not well understood. In folded proteins, as in protein-ligand complexes, the consideration of hydrogen bonding is complicated by the requirement to take into account the role of solvent in stabilizing both the folded and the unfolded state (or, in the case of protein-ligand complexes, the associated and dissociated states).

The role of hydrogen bonding at one site in T4 lysozyme has been analyzed by a series of substitutions of Thr 157. In wild-type lysozyme the γ-hydroxyl of this threonine participates in a network of hydrogen bonds (Figure 2). Early studies of randomly generated temperature-sensitive mutant lysozymes had shown that the replacement of threonine with isoleucine at position 157 substantially destabilizes the protein, apparently because it results in the disruption of the hydrogen-bond network (Alber et al., 1986, 1987; Grütter et al., 1987). Other changes between the two structures could, however, also contribute to instability. To determine how Thr 157 contributes to the stability of T4 lysozyme, 13 different amino acids were substituted at this site. The structures of these modified lysozymes have been determined and their stabilities measured (Alber et al., 1987). The results show that the main way in which Thr 157 contributes to stability is through its hydrogen-bonding interactions. An interesting situation occurs when glycine is substituted at position 157. The lack of a side chain allows a water molecule to bind at the site previously occupied by the γ-hydroxyl of the threonine and to restore the hydrogen-bond network, giving a protein whose stability is close to wild-type (Figure 2).

In the case of the 13 different substitutions that have been made at position 157 in T4 lysozyme, no mutant is more stable than wild-type. Also, no engineered mutant protein is of lower stability than Ile 157, the variant that was obtained as a temperature-sensitive mutant after random chemical mutagenesis. This tends to suggest that proteins are tolerant of change and, within reason, relatively resistant to destabilization by amino acid replacements.

The above study suggests that one way in which proteins might be stabilized would be to locate any hydrogen bond donors or acceptors in the folded structure of the protein that do not participate in hydrogen bonding and to satisfy their H-bonding potential by appropriate site-directed substitutions. A detailed analysis of the T4 lysozyme structure, using the program of Baker and Hubbard (1984), did not reveal a single candidate for such a substitution. Within the accuracy of the X-ray structural analysis it appears that every hydrogen bond donor and acceptor in T4 lysozyme participates in at least one hydrogen bond (although not necessarily with good geometry). Unsatisfied H-bonding groups in folded proteins seem to be very rare, perhaps because they would tend to be very destabilizing, and the protein structure therefore relaxes to alternative conformations which permit hydrogen bonds to occur.

Helix-dipole interactions

Recent evidence has shown that the stabilities of proteins can be enhanced by the introduction of appropriately charged groups at the ends of α-helices (Mitchinson and Baldwin, 1986; Nicholson, Becktel and Matthews, 1988).

In the case of T4 lysozyme initial experiments have focused on the introduction of aspartic acids at or near the amino termini of α-helices. Two such substitutions, Ser 38 → Asp and Asn 144 → Asp, were both found to increase the melting temperature of the protein by about 2 °C at pH values where the introduced aspartates were negatively charged. The double mutant was found to increase the melting temperature by about 4 °C (Nicholson et al., 1988). A related substitution, Asn 144 → Glu has also been constructed and found to yield essentially the same increase in stability as the replacement with aspartic acid (Nicholson et al., unpublished observations). This therefore seems to be a rather general way to increase protein stability.

Structural studies of the wild-type and mutant lysozymes indicate that the stabilization is due to generalized electrostatic interaction of the introduced aspartic acid side chain with the positive charge at the end of the α-helix, and does not require precise hydrogen bonding to the terminal amino groups. In the case of the Asn 144 → Asp substitution, for example, neither the Asn or Asp side chain makes any hydrogen bonds to the end of the helix. Because precise hydrogen bonding is not required it greatly simplifies the design of stabilizing substitutions.

Substitutions that decrease the entropy of unfolding

It has been proposed that selected substitutions of the form Xaa → Pro and Gly → Xaa can be used to decrease the configurational entropy of unfolding of a protein and so increase its thermostability (Matthews, Nicholson and Becktel, 1987). The basic idea is that glycine has greater conformational flexibility than a residue with a β-carbon and so requires greater free energy to change from the unfolded to the folded state. Conversely, proline has a very restricted conformation and so requires a relatively low expenditure of free energy to retain in the folded state. The sites in a protein at which glycines are to be replaced or prolines substituted must, of course, be chosen in such a way that the native structure of the protein is not perturbed.

One such substitution, Ala 82 → Pro was found to increase the melting temperature of T4 lysozyme by 2°C (Matthews et al., 1987). In this case a crystal structure analysis confirmed that the structure of the mutant protein was virtually identical with that of wild-type, apart from the addition of the pyrrolidine ring of the proline. A second substitution, Ala 93 → Pro, enhanced the stability of T4 lysozyme toward irreversible denaturation, but increased the reversible melting temperature at pH 6.5 by only 0.2°C (Nicholson et al., unpublished observations).

The replacement of Gly 77 with alanine was found to increase the melting temperature by 1 °C at pH 6.5 (although not at pH 2.0) (Matthews et al., 1987). A second replacement, Gly 113 → Ala increased the melting temperature by 0.6 °C at pH 5.0 and 0.3 °C at pH 2.0 (Nicholson et al., unpublished observations).

Thus the removal of glycine residues and the insertion of prolines can enhance thermostability, although the increase in melting temperature is only marginal in some cases.

The above approach is not restricted to substitutions involving glycines and prolines. Replacements of the form Ser → Thr and Ala → Val, among others, can also be considered. For example, the variant Ala 41 → Val has been constructed and found to increase the melting temperature by 1.9 °C (Dao-pin et al., unpublished results).

Removal of strain

A possible approach to stabilizing proteins might be to identify parts of the protein structure that are under strain and to relieve such strain by appropriate amino acid replacements. One difficulty in such an approach is that strain is likely to be distributed over the whole protein and not localized at a few sites. Also, the distortions due to strain that might be anticipated in a typical protein structure are likely to be difficult to detect, given the relatively limited accuracy that can be achieved even from highly refined protein crystal structures.

One approach along these lines that has been tested in T4 lysozyme is to replace so-called "left-handed helical" residues with glycine (Figure 3). The rationale is that non-glycine residues are rarely observed to have conformations on the right-hand side of the Ramachandran

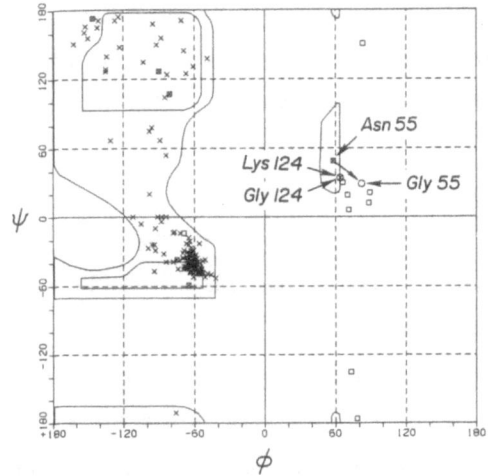

Figure 3. Ramachandran diagram of T4 lysozyme showing the (ϕ, ψ) conformations of the two "left-handed-helical" residues, Asn 55 and Lys 124, that were respectively replaced with glycine. In neither case is the stability of the protein significantly altered.

diagram, whereas such conformations for glycines are common and are expected to be of lower energy.

Replacements of both Lys 124 and Asn 55 with glycine were constructed and found to have free energies of folding virtually identical with that of wild-type lysozyme (Nicholson et al., 1989). This indicates that there is actually very little difference between the energy of a glycine and a non-glycine in the "left-handed helical" conformation.

Another approach to the removal of strain is to look for residues that have values of the side-chain rotation angle χ^1 that are displaced from the expected energy minima. One possible candidate in T4 lysozyme is Val 131, for which the χ^1 value is 18° from the low-energy value. Val 131 is within an α-helix and the rotation of the side chain is presumably due to a close contact with atoms in the next turn of the helix. Replacement of Val 131 with alanine increased the melting temperature of the protein by 1.1 °C (Dao-pin et al., 1989). In this case the design was successful although the X-ray crystallographic structure analysis suggests that the expected stabilization due to relaxation of the x^1 angle is offset somewhat by loss of hydrophobic stabilization and by entropic stabilization associated with the Val → Ala replacement.

Disulfide bridges

A number of attempts have been made to increase the stability of proteins by the introduction of non-native disulfide bridges (Villafranca et al., 1983, 1987; Perry and Wetzel, 1984; Sauer et al., 1986; Bryan et al., 1985; Wells and Powers, 1986). Surveys of disulfide bridges in known protein structures show that the geometry of such bridges is very restricted (Thornton, 1981; Richardson, 1981; Pabo and Suchanek, 1986). For this reason it is often difficult to find suitable pairs of residues in a protein that can be linked by a disulfide bridge without concomitant introduction of strain. Although the polymer theory indicates that genetically engineered disulfide bridges can increase protein stability, experience to date shows that in many cases the engineered protein is not more stable than wild type.

In an attempt to elucidate general principles relevant to the design of disulfide linkages, five different bridges have been introduced into phage lysozyme (Figure 4) (Perry and Wetzel, 1984; Matsumura et al., 1989a). Three of these bridges increase the melting temperature of the protein by 5-11 °C at pH 2; the other two bridges destabilize the protein relative to wild-type lysozyme (Figure 5). In each case the oxidized

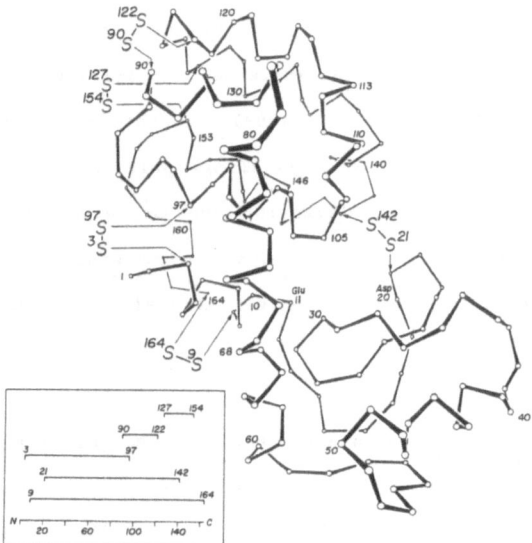

Figure 4. Locations of five disulfide bridges that have been engineered into T4 lysozyme. The lengths of the loops formed by these bridges are shown schematically in the insert.

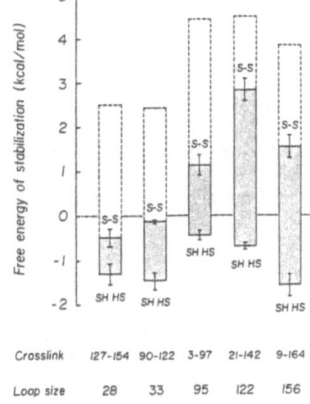

Figure 5. Schematic diagram showing the free energies of stabilization of the reduced (SH) and oxidized (SS) forms of the disulfide-bridge mutant lysozymes shown in Figure 4. The dotted line illustrates the theoretical stabilization to be expected from the reduction in entropy of the unfolded form, calculated using the formula of Pace et al. (1988). This theoretical contribution to the free energy of stabilization is added to the observed free energy of the reduced form. (Based on Matsumura et al., 1989a.)

(crosslinked) form of the protein is more stable than the reduced (noncrosslinked) form.

It is anticipated that stabilization from disulfide bridges arises from the reduction in entropy of the unfolded protein and increases logarithmically with the size of the loop that is formed (e.g. see Pace et al., 1988). In practice, this contribution to stability will be offset by loss of pre-existing favorable interactions resulting from the replacement of residues with cysteines and/or by strain associated with the formation of the S-S bridge.

As seen in Figure 5 the results obtained with T4 lysozyme are consistent with these expectations. In each case the reduced mutant proteins are about 0.5-1.5 kcal/mol less stable than wild-type, suggesting that the introduction of each single cysteine decreased the stability of the protein by about 0.75 kcal/mole on average. In no case is the stabilization expected from the entropic effect fully realized. Rather, the formation of each disulfide bridge appears to introduce strain into the molecule that reduces the net stability of the oxidized form of the mutant protein. This inference is supported by the observation that the disulfide bridges that appear to be under greatest strain (127-154 and 90-122) have measured stabilities substantially lower than the theoretical upper limit (Matsumura et al., 1989a).

The three disulfide bridges that are most effective in stabilizing T4 lysozyme have the largest loop sizes. Consistent with theoretical expectation, this suggests that a large loop size is a desirable attribute of engineered disulfide bridges that are intended to maximize protein stability.

When possible, disulfide bridges should also be introduced at sites at which the conformation of the native protein is geometrically

92

compatible with the known requirements for formation of an unstrained disulfide bridge. Bearing in mind, however, that such sites are rare, it may be necessary to choose a site that is less than ideal. In the case of T4 lysozyme, the two bridges that are most effective in stabilizing the protein (9-164 and 21-142) are introduced into a flexible part of the structure. The use of such flexible sites could be another general attribute that is desirable in designing disulfide bridges in general (Matsumura et al., 1989a).

CONCLUSIONS

There are many possible ways in which the thermostability of proteins might be increased.

Tests to date with T4 lysozyme suggest that the most effective approaches are: (1) use of disulfide bridges, (2) stabilization of α-helix dipoles, and (3) entropic stabilization via the introduction of prolines and/or removal of glycines.

It also appears that the effects of independent mutations are additive. The combination of different mutations, each of which may provide only a modest increase in thermostability, therefore holds the promise of engineering proteins of substantially enhanced thermostability.

ACKNOWLEDGEMENTS

We have benefitted by receiving help and advice from a number of collaborators in the Institute of Molecular Biology including W. Baase, W.J. Becktel, F.W. Dahlquist, M. Lindorfer, D.C. Muchmore, S.J. Remington and J.A. Schellman.

The work was supported in part by grants from the National Institutes of Health (GM21967; GM20066), the National Science Foundation (DMB8611084) and the Lucille P. Markey Charitable Trust.

REFERENCES

Alber, T., Dao-pin, S., Wilson, K. Wozniak, J.A., Cook, S.P. and Matthews, B.W. (1987) Nature <u>330</u>, 41-46.

Alber, T., Grütter, M.G., Gray, T.M., Wozniak, J., Weaver, L.H., Chen, B-L., Baker, E.N. and Matthews, B.W. (1986) UCLA Symposium on Molecular and Cellular Biology, New Series, D.L. Oxender, ed. Alan R. Liss, Inc., NY <u>39</u>, 307-318.

Baker, E.N. and Hubbard, R.E. (1984) Prog. Biophys. Molec. Biol. <u>44</u>, 97-179.

Bell, J.A., Dao-pin, S., Faber, R., Jacobson, R., Karpusas, M., Matsumura, M., Nicholson, H., Pjura, P.E., Tronrud, D.E., Weaver, L.H., Wilson, K., Wozniak, J.A., Zhang, X-J., Alber, T. and Matthews, B.W. (1989) In "The Use of Crystallography in the Design of Antiviral Agents", W.G. Laver and G.M. Air, eds. Academic Press, Orlando, Florida.

Bryan, P., Rollence, M., Pantoliano, M., Quill, S., Wood, J., Perna, L., Matthew, J., Hsiao, H. and Poulos, T. (1985) J. Cell. Biochem. Suppl. <u>9B</u>, 92.

Dao-pin, S., Baase, W. and Matthews, B.W. (1989) Proteins: Structure, Function and Genetics, manuscript submitted.

Grütter, M.G, Gray, T.M., Weaver, L.H., Alber, T., Wilson, K. and Matthews, B.W. (1987) J. Mol. Biol. <u>197</u>, 315-329.

Kauzmann, W. (1959) Adv. Prot. Chem. <u>14</u>, 1-63.

Matsumura, M., Becktel, W.J. and Matthews, B.W. (1988) Nature <u>334</u>, 406-410.

Matsumura, M., Becktel, W.J., Levitt, M. and Matthews, B.W. (1989a) Proc. Natl. Acad. Sci. USA, in press.

Matsumura, M., Wozniak, J.A., Daopin, S. and Matthews, B.W. (1989b) J. Biol. Chem., manuscript submitted.

Matthews, B.W. (1987) Biochemistry <u>26</u>, 6885-6888.

Matthews, B.W., Nicholson, H. and Becktel, W.J. (1987) Proc. Natl. Acad. Sci. USA <u>84</u>, 6663-6667.

94

Mitchinson, C. and Baldwin, R.L. (1986) Proteins: Struct. Funct. Genet. 1, 23-33.

Nicholson, H., Becktel, W.J. and Matthews, B.W. (1988) Nature 336, 651-656.

Nicholson, H., Söderlind, E., Tronrud, D.E. and Matthews, B.W. (1989) J. Mol. Biol., submitted.

Pabo, C.O. and Suchanek, E.G. (1986) Biochemistry 25, 5987-5991.

Pace, C.N., Grimsley, G.R., Thomson, J.A. and Barnett, B.J. (1988) J. Biol. Chem. 263, 11820-11825.

Pantoliano, M.W., Ladner, R.C., Bryan, P.N., Rollence, M.L., Wood, J.F. and Poulos, T.L. (1987) Biochemistry 26, 2077-2082.

Perry, L.J. and Wetzel, R. (1984) Science 226, 555-557.

Richardson, J.S. (1981) Adv. Prot. Chem. 34, 167-339.

Sauer, R.T., Hehir, K., Stearman, R.S., Weiss, M.A., Jeitler-Nilsson, A., Suchanek, E.G. and Pabo, C.O. (1986) Biochemistry 25, 5992-5998.

Thornton, J.M. (1981) J. Mol. Biol. 151, 261-287.

Villafranca, J.E., Howell, E.E., Oatley, S.J., Xuong, N-H. and Kraut, J. (1987) Biochemistry 26, 2182-2189.

Wells, J.A. and Powers, D.B. (1986) J. Biol. Chem. 261, 6564-6570.

Stability of Folded Conformations by Computer Simulation: Methods and Some Applications

JAN HERMANS, R.-H. YUN, and AMIL G. ANDERSON

Molecular Dynamics Calculations

Molecular dynamics calculations have rapidly developed into a versatile technique for studying molecular conformation, dynamics and equilibria of macromolecules. Molecular motion is modeled with classical mechanics, using a simple forcefield in which the potential energy is calculated as a sum of terms each of which depends on the coordinates of only a few atoms. Solvation is represented most accurately by including in the sample sufficient solvent molecules to entirely envelop the solute.

As the calculation progresses, the system will traverse many different states, approaching an increasingly perfect sampling of the Boltzmann distribution, i.e., of the thermal-equilibrium ensemble of states. By application of statistical mechanics one can obtain ensemble averages, including thermodynamic functions such as energy, enthalpy, specific heat and free energy. The calculation of conformational distributions is, in nearly all cases of interest, complicated by the existence of many conformations for which the free energy is a (local) minimum, separated by free energy barriers of different height. The barriers are rarely crossed sufficiently frequently for the establishment of a statistically significant distribution over several distinct conformations in the limited time represented in a computer simulation. The difficulties of obtaining statistical sampling of different conformational states can, however, be circumvented by the use of conformational forcing.

Conformational forcing is one of several free energy simulation methods.(Berendsen et al. 1985; Beveridge & Mezei 1985; Beveridge 1986; van Gunsteren & Weiner 1989) Free energy simulations require artificial modification of the forcefield in order to drive the system from one state to another; frequently the system passes through a series of intermediate states that are physically unrealizable. In the case of conformational forcing, the system is perturbed with a term supplying a torque that is sufficiently strong to drive the system over the energy barriers (Anderson et al. 1986). Another example of a free energy simulation is the "molecular replacement" calculation, which estimates the thermodynamics of changing one molecule into another by gradually changing the forcefield (Warshel et al., 1986; Hwang et al., 1987; McCammon et al. 1986; Wong & McCammon 1986, 1987; Rao et al., 1987).

We summarize here the results of our applications of free energy simulations, first, to the calculation of conformational equilibria and, second, to the problem of the *change* of stability of a folded conformation upon amino acid substitution.

Calculation of Conformational Probability Distributions

As mentioned, a complete conformational distribution for a peptide cannot be obtained in a single free, i.e., unperturbed, dynamics simulation: A free simulation will often be restricted to the neighborhood of a single conformational free energy minimum; in addition, states with somewhat higher free energy will not be sampled reliably. In order to calculate a complete conformational probability distribution, and the associated free energy map, we have proceeded by first calculating non-overlapping, local distributions in the neighborhood of all free energy minima. Then free energy differences between these states have been calculated by free energy simulation, as described next. With a knowledge of local conformational probability distributions, P_i, and their relative standard free energies, ΔG_i° the global probability distribution could then be calculated with

$$P_0(\chi) = \sum_i P_i(\chi) \exp(-\Delta G_i^\circ/kT) \qquad [1]$$

and subsequent normalization. Conformational differences correspond in first instance to differences in internal rotation about single bonds in the molecule, i.e., to changes in dihedral angles, χ (in peptides, backbone dihedral angles ϕ and ψ, and side chain dihedral angles).

Free energy calculations

Free energy calculations, i.e., molecular dynamics simulations in which free energy differences are calculated, are done with use of a varying potential. Sometimes one employs a forcing potential, which is extrinsic to, and added to, the potential energy terms that describe the molecular physics of bonded and nonbonded interactions. The variation of the potential during the calculation drives the system from one state to another, for example, from one conformation to another. In the process, the forcing potential performs work in order to change the system. If the change is made slowly enough, then, by definition, the work performed by the forcing potential on the system is equal to the change of the *free* energy of the system. Such a process is called quasi-static. Thus,

$$\Delta A_{AB} \text{ or } \Delta G_{AB} = \int dW \quad \text{(quasi-static process)} \qquad [2]$$

The Helmholtz or Gibbs free energy may be used, the latter if the process takes place at constant pressure, which is usually the case for both experiment and simulation. The forcing potential, U_f is written as a function of a coupling parameter, λ, such that when $\lambda=0$ the system is in state A, and when $\lambda=1$, the system is in state B. The work for the change of state is computed as

$$\Delta G_{AB} = \int_0^1 <\partial U_f/\partial \lambda> \, d\lambda \qquad [3]$$

The $<>$ sign indicates the quasi-static nature of the process, which means that the integrand must be averaged over a Boltzmann distribution. In practice, free energy differences can be computed in molecular dynamics simulations in which λ, and hence the forcing potential, changes very slowly. (Some authors prefer an alternate, closely related method of computing the free energy difference with stepwise increments of λ. The two methods are found to give similar precision for similar expenditures of computer time.)

Conformational forcing.

In order to produce conformation change by internal rotation about single bonds we have used forcing potentials of the form

$$U_f = (K_f/2) \, [1 - \cos(\chi - \chi_0)] \qquad [4]$$

Such a potential restrains a dihedral angle χ to the neighborhood of χ_0. The value of χ_0 can be changed and coupled to a parameter λ with

$$\chi_0 = \chi_B + (1-\lambda) \, (\chi_A - \chi_B) \qquad [5]$$

A free energy difference calculated with use of a forcing potential of this form, pertains to states with an artificial restraint. The free energy difference for the unrestrained states is obtained by adding the difference between the free energies for applying the restraint in states A and B. The restraint free energy for state A is easily calculated with

$$\Delta G_{f,A} = -kT \ln <\exp(-U_f/kT)>_A = -kT \ln [\int d\chi \, P_A(\chi) \exp(-U_f/kT)] \qquad [6]$$

where $P_A(\chi)$ is the conformational probability distribution for state A, and similarly for state B.

Molecular replacement.

In a molecular replacement simulation, the potential energy function is (gradually) changed from that describing one molecule (Q) to that for another molecule (R). The system's potential energy, U_{pot} is written as a function of a coupling parameter, λ, in such a way that

$$U_{pot}(0) = U_Q \text{ and } U_{pot}(1) = U_R \qquad [7]$$

In its simplest form,

$$U_{pot} = U_Q - \lambda(U_Q - U_R) \qquad [8]$$

The conformational free energy difference for converting molecule Q into molecule R may then be found with

$$\Delta G_{QR}° = \int_0^1 <\partial U_{pot}/\partial \lambda> d\lambda \qquad [9]$$

As an alternative, it is frequently possible to use molecular replacement calculations in order to obtain conformational free energy differences. We use as an example the problem of calculating the conformational distribution for internal rotation about the $C_\alpha - C_\beta$ bond of a valine side chain in a peptide or protein molecule. One can expect this distribution to have three maxima, near $\chi = 60°$, $-60°$ and $180°$. The free energy differences between these conformations can be established by conformational forcing. Alternatively, one can compute free energies for molecular replacement of alanine by valine in each of the three conformations, and calculate the conformational free energy differences by subtraction. In each calculation, the valine side chain is restrained to the neighborhood of one of these conformations, with a *constant* forcing potential. We have shown that this produces the same answer as conformational forcing; if the barriers for internal rotation are large, the approach *via* molecular replacement is the easier one (Hermans et al. 1989).

Applications to Peptide Conformation

Alanine dipeptide

The alanine dipeptide, N-acetyl-alanyl-N-methylamide, serves as a paradigm for studying the thermodynamics of protein conformation and folding. The dipeptide has two major degrees of freedom, the dihedral angles ϕ, for rotation about the N-Cα bond, and ψ, for rotation about the Cα-C bond, which also determine the conformation of the carbon backbone of proteins. We have computed the conformational equilibrium distribution of the dipeptide in water. In addition, we have mapped the free energy along the pathways between conformations.

Two complementary approaches were taken to calculate the ϕ-ψ conformational probability surface for the alanine dipeptide. First, separate simulations were run in the regions of four previously identified potential energy minima conformations. The probability surfaces derived from the trajectories of these simulations identified the most probable conformations for our model of the dipeptide in solution. The free energy differences between the four energy minima were then calculated using conformational forcing. With these free energy differences we could scale together the regional probability distributions to produce an overall ϕ-ψ probability map.

One can see from the map in Figure 1 that the ϕ-ψ conformations representing the free energy minima are $\beta(-110°,120°)$, $\alpha_R(-120°,-40°)$, $\alpha_L(60°,100°)$ and $C^7_{ax}(70°,-60°)$. In first approximation, the conformational distribution of the hydrated dipeptide represents that of a residue in an unfolded protein or polypeptide. This is clearly dominated by the β conformation (90% of the equilibrium population at room temperature). Any cycle of four successive conformation changes, which returns to the starting conformation has a closure error of less than 1 kJ/mole, a value that greatly exceeds the precision of the currently available experimental data for the solution equilibria of the alanine dipeptide (Avignon et al. 1973; Madison & Kopple 1980). Thus, the internal consistency and precision of our results are excellent. Recent advances

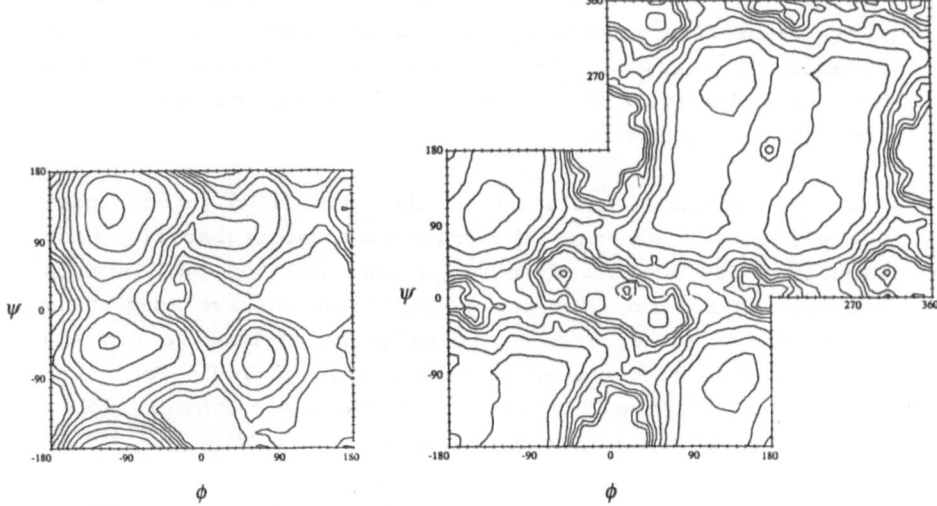

Figure 1. Conformational free energy maps of the alanine and glycine dipeptides; contours are at 2 kJ/mole intervals. All values are positive, with the value of zero corresponding to the lowest minima (for alanine at ϕ=130°, ψ=-110°; for glycine at ϕ=110°, ψ=-130° and ϕ=-110°, ψ=130°).

in peptide synthesis and NMR may allow experimental verification in the near future.

The free energy barriers along the transition paths between the different conformations vary from 1 to 10 kJ/mole above the free energy difference between end point conformations. The elementary step of the folding process, i.e. internal rotation about a main chain single bond, can apparently occur with ease about both bonds on either side of a small side chain.

Glycine Dipeptide

The glycine dipeptide provides an interesting contrast with the alanine dipeptide. Replacing the methyl group with a hydrogen leads to a molecule with much greater conformational freedom. This was noted in the earliest steric clash calculations by Ramachandran. The β and α_R regions are quite similar in energy and because of the the molecular symmetry, β and

C_7^{ax} have the same energies, as do α_R and α_L. Chou & Fasman (1974) classified glycine as an α-helix breaker, that is, the presence of a glycine residue within a sequence of amino acids was negatively correlated with the presence of helical structure in a protein.

A long unperturbed simulation of the hydrated glycine dipeptide was used to generate the conformational probability distribution: 500,000 steps corresponding to 1000 ps. In this distribution all four low-energy conformations are adequately represented, the barriers to rotation being between 0.5 and 1 kJ/mole when either ϕ or ψ changes through 180°. For rotation through 0°, the barriers are higher, 6.5 kJ/mole for a change of ψ, with ϕ in the neighborhood of −110°.

A preliminary analysis of the kinetics of conformation change of the hydrated glycine dipeptide is possible as a consequence of the unforced exploration of ϕ,ψ conformation space. In the course of the 1000 ps simulation there were 97 transitions between the four distinguishable regions of large stability (cf. Figure 1). Trajectories during these transitions appear to be diffusive in nature, i.e., the motion is strongly damped by friction with solvent. When compared with the free energy maps, these results make qualitative sense in terms of barrier crossings and path lengths needed for the different transitions.

α-Helix: Terminal Residues

We have made considerable progress with a molecular dynamics free energy study of a solvated deca-L-alanine α-helix. The objective of this study is to determine the conformational probability distribution of the terminal residues, in particular the equilibrium between the extended β conformation and the hydrogen-bonded α_R conformation of the terminal residues. (The two ends of the helix were examined separately as they are not chemically identical.) This conformation change extends an existing helix by one residue and, except for possible effects due to the location of the residues at the ends of both chain and helix, the equilibrium constant corresponds to the propagation parameter, s in the theory of the helix-coil transition. The value of s is from experiment known to be close to 1.0, i.e., quite different from the equilibrium constant for $\beta \leftrightarrow \alpha_R$ (0.1) that we had calculated for the model of the alanine dipeptide.

In all systems, residues that were to remain in the helical conformation were restrained so as to maintain a core helical structure. The free energy for unwinding one residue at either end of the helix was calculated using a dihedral angle forcing potential. For the C-terminal end of the helix, reversibility of the calculations was good; the free energy change for extending the helix by one residue was approximately 3 kJ/mole, which corresponds to an equilibrium constant, s of about 3.

For the N-terminal residue reversibility was very poor. Examination of the trajectories with graphics showed that the last intramolecular hydrogen bond at the N-terminus would break (or reform) abruptly, and only once per simulation. A more precise analysis of the trajectories verified this. Thus, a significant strain built up in the intramolecular hydrogen bond in the peptide, which was released in an irreversible fashion when hydrogen bonds with solvent water formed late in the simulation. In contrast, the breaking of the hydrogen bond and formation of hydrogen bonds with solvent molecules in the process of unwinding the C-terminal residue took place much more smoothly. Examination via molecular graphics suggests that the N-terminal hydrogen bond is less solvent-accessible than the C-terminal one, due to shielding by the methyl groups.

The free energy change for the helix$\rightarrow$coil transition, while closer to zero than that for $\alpha_R \rightarrow \beta$ transition in the alanine dipeptide, still corresponds to that for an unstable helix. We expect that an increase of the cutoff distance beyond 8 Å will increase the calculated stability of the model hydrated helix, by bringing additional, presumably stabilizing, dipole-dipole interactions between peptide groups in the helix into play. As an increase of R_c increases the computing effort by a factor of $R_c{}^6$, full molecular dynamics simulations at larger cutoffs somewhat exceeded our computing resources.

Because the number of distinct conformations of a peptide increases exponentially with the number of single bonds, it is clear that the scope of these calculations with very detailed representation of atomic interactions, including solvation, is very limited. We can expect to slowly increase the complexity of the systems under study, particularly as computer resources can be expected to grow. In the following part of this paper we discuss how the scope of these calculations can be broadened, even with use of present resources, to include studies of the effect of amino substitution on the free energy of unfolding of any particular stable conformation of a peptide (such as the α-helix), or the native conformation of a protein.

Conformational Stability and Amino Acid Substitution

Much new experimental information is available about how the stability of folded conformations changes due to single amino acid substitutions, thanks to site-specific mutagenesis of proteins and solid-phase synthesis of oligopeptides. This new data base has led to speculation about how these changes can be attributed to different physical effects known to influence conformational stability. One must consider the importance of changes in a variety of structural parameters: hydrogen bonding, charge-charge and charge-dipole interactions, hydrophobic contacts, bad steric contacts and extent of conformational freedom, as well as a possible overall conformation change as a result of the substitution. This classification and quantification, while very interesting, has been until now largely empirical.

We have used molecular dynamics simulations to obtain a quantitative and detailed analysis of changes in conformational stability in three test structures in which alanine was replaced with glycine.

The change in stability of a peptide conformation due to amino acid substitution is generally expressed as the change in standard free energy of unfolding, $\Delta\Delta G°$. It has been recognized for some time that molecular dynamics simulation with applied forcing potentials can be used to make theoretical estimates of $\Delta\Delta G°$ (Berendsen et al. 1985). One applies the following thermodynamic cycle,

$$
\begin{array}{ccccc}
(1) & F(Ala\text{-}i) & \Leftrightarrow & U(Ala\text{-}i) & (2) \\
& \Uparrow & & \Uparrow & \\
& \Downarrow & & \Downarrow & \qquad [10] \\
(3) & F(Gly\text{-}i) & \Leftrightarrow & U(Gly\text{-}i) & (4)
\end{array}
$$

in which F and U represent folded and unfolded conformations. Ala-i and Gly-i are two variants of a protein, one with alanine, the other with glycine in position i. The horizontal steps represent the two unfolding equilibria, which can be studied experimentally. The desired free energy increment is

$$\Delta\Delta G° = \Delta G°_{12} - \Delta G°_{34} = \Delta G°_{13} - \Delta G°_{24} \qquad [11]$$

The free energy differences $\Delta G°_{13}$ and $\Delta G°_{24}$ are those for substitution of alanine with glycine at position i in the two conformations, folded and unfolded; these differences may be estimated theoretically *via* molecular dynamics simulation. Methods for such molecular substitution calculations have been developed and tested both in systems of small molecules (Bash et al. 1987a; Straatsma et al. 1986) and for protein-small molecule complexes, (Bash et al. 1987b; Hermans & Shankar 1987) with remarkably good agreement with experiment.

We used, in addition, several methods to deal with changes of conformation and conformational flexibility: conformational restraints and constraints (van Gunsteren & Berendsen 1985) were applied, respectively, to prevent and to obtain conformation change, and to calculate corresponding free energy differences (Anderson et al. 1986; Anderson & Hermans 1988). These were particularly important in dealing with the inherently very flexible random-coil conformation.

Random-Coil

We make a crucial assumption about linear random polypeptide chains, which is that the conformational distribution of each residue, as characterized by the internal rotation about the main-chain bonds on either side of the α-carbon atom and about single bonds in the side chain, is independent of the conformation of all other residues in the molecule (Brant & Flory 1965; Brant et al. 1967). Accordingly, the conformational free energy in the random coil state can be found by using the conformational distribution of small oligopeptides, the (alanine and glycine) dipeptides. The free energy difference for replacing alanine by glycine in the random coil, i.e., $\Delta G°_{24}$, contains contributions for adding the side chain methyl group and for the difference in conformational free energy. This free energy difference was calculated as 1.5 kJ/mole. This small net difference hides a significant term of -4.5 kJ/mole due to the large difference in conformational freedom of the alanine and glycine backbones, illustrated in Figure 1. This value of $\Delta G°_{24}$ is applicable to *all* cases where a stability change is caused by substitution of alanine with glycine. That is, thanks to the assumption of independence of residue conformation in the random coil state, the right hand side of scheme 1 is *de facto* always the same.

Helix

The stability of α-helices of homopolypeptides is adequately described by the theory of the helix-coil transition in terms of the equilibrium constants for helix growth and initiation, s and σ (Zimm & Bragg 1959). The stability of a helix n residues long is given by its equilibrium constant relative to the random coil state; the value of this equilibrium constant is σs^n. For alanine and glycine the constants σ and s have been measured by Scheraga and coworkers using the host-guest technique and copolymers with several hundred residues. All reported experimental results indicate that alanine forms more stable helices than glycine; this correlates with the higher relative occurrence of alanine in α-helices of globular proteins. Reported values at 30°C are: $s_{Ala}=1.07$ and $s_{Gly}=0.59$ (Sueki et al. 1984). Since in the simulations a glycine residue was substituted in the middle of an alanine helix and the end residues remained alanines, the calculated difference in stability is attributable to a difference in helix *growth* parameter, which gives $\Delta\Delta G° = -RT \ln (s_{Gly}/s_{Ala}) = 1.48$ kJ/mole. A second experimental estimate has been obtained recently for a synthetic oligo-peptide which can form an amphiphilic helix, two of which associate *via* their hydrophobic sides. From precise measurements of the equilibrium between helical dimer and random coil monomer in urea solutions follows an experimental value of $\Delta\Delta G°=4.5$ kJ/mole per replaced residue (O'Neill & DeGrado, unpublished results). A third estimate follows from recently measured differences in helix stability for synthetic C-peptide and analogs (Strehlow & Baldwin, 1989). The estimate derived from this work is in the range of the other two.

The simulation gives a net difference $\Delta\Delta G°$ of 3.4 kJ/mole. The greater conformational freedom of the glycine residue has in the past been recognized as contributing significantly to the destabilization of folded conformations when glycine substitutions are made. As a result of being done in four steps, i.e., two calculations of replacement free energy and two calculations of conformational free energy, this simulation confirms this quantitatively: approximately the same free energy change is calculated for adding a methyl group to α-helical and (restrained) extended molecules, whereas the conformational free energies of the unfolded glycine and alanine peptides differ by almost 5 kJ/mole.

We performed three different replacement calculations: in the first a central alanine in an alanine oligomer was replaced by glycine, in the second a central alanine residue in a glycine oligomer was replaced by glycine and in the third all

ten residues of the alanine oligomer were replaced by
glycine. The results of these calculations indicate that the
free energy contribution per replaced residue is independent
of sequence and position in sequence, an assumption that frequently underlies application of the theory of the helix-coil
transition, including the interpretation of the cited host-
guest experiments. (This result does not necessarily extend
to any residues other than alanine and glycine.) According
to the simulation results, replacement of end residues
changes the stability by the same amount as replacement of
central residues.[1]

β-Turn

Evidence that some fragments of native protein molecules in
isolation have a significant preference for the conformation
they acquire when part of the folded protein, had been available for some time (Scheraga & Epand, 1968; Hermans & Puett,
1971). More recently, synthesis of oligopeptides of arbitrary
sequence by solid-phase methods has produced striking examples of model peptides that readily acquire secondary structure, such as helices (Marqusee et al. 1987) and turns. An
example is the pentapeptide YPXDV, which has been studied as
a series of 18 variants (Dyson et al. 1988; Wright et al.
1988). Characterization by two-dimensional nmr has shown the
glycine-variant, i.e., YPGDV, to form a significant percentage of β-turn structure, and all other variants to have a
lesser tendency to form the β-turn conformation, with the alanine variant showing the least, and possibly no observable,
tendency.

Our simulations clearly indicate that the glycine variant
should indeed form the more stable structure. The calculated

---

[1] Accordingly, our results do not suggest a difference in the
helix initiation parameters, σ for alanine and glycine as
large as that determined experimentally by the host-guest
technique, $(\sigma_{Ala}/\sigma_{Gly})=80$ (Sueki et al. 1984). However, it
should be noted that experimental σ-values are much less
precisely known than experimental s-values.

free energy difference, $\Delta\Delta G^{o}$ of 20 kJ/mole far exceeds the minimum value set by the experimental results. The difference is this large when the molecule assumes a Type II β-turn. These two types of turn were found to have approximately the same free energy in the alanine derivative, but in the glycine derivative the Type II turn has a much lower free energy than the Type I turn. A simple rationale for the large differences (Rose et al. 1985; Wilmot & Thornton 1988) considers the conformation of the proline and glycine (or alanine) residues in the turn. In the type II turn conformation, the proline residue in the first turn position has a low-energy conformation, while in the second turn position an alanine residue has a high- and a glycine residue a low-energy conformation. In the type I turn, both turn residues are constrained to moderately higher-energy conformations, with alanine and glycine about equal (Figure 2; *cf.* free energy maps for the dipeptides in Figure 1).

It is therefore not surprising that glycine is a highly preferred residue in the second turn position in globular proteins and that β-turns with glycine in this position are overwhelmingly type II turns: this corresponds to a great difference in stability of the isolated type I and type II turns containing glycine. Thus, the presence of glycine will favor formation of type II turns during the protein folding process. On the other hand, turns with alanine will not form

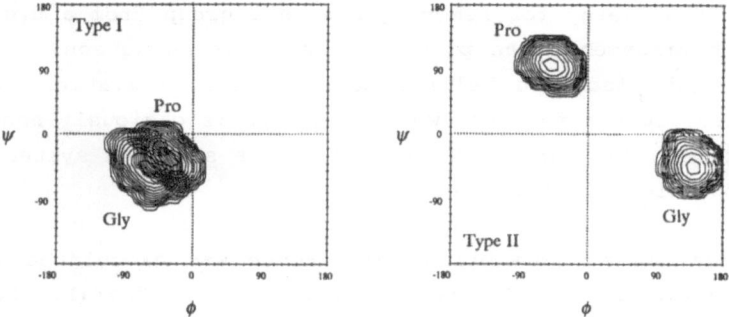

Figure 2. Conformational distribution of proline and glycine residues in the Type I and II β-turn model, from dynamics simulations.

nearly as readily, *and* will not have a strong preference for either type I or type II.

T4-Lysozyme

Many stability mutants of T4-lysozyme have been characterized (Matthews et al. 1987; Alber et al. 1987, 1988; Gray et al. 1987; Matsumura et al. 1988). For the Ala146Gly mutant a refined crystal structure shows an absence of overall conformation change relative to the wild-type protein (B.W. Matthews, unpublished results). The mutant is less stable, i.e., unfolds at a lower temperature. Residue 146 is part of an α-helix. The methyl group is completely buried: it is surrounded by two hydrogen-bonded backbone NH and CO groups, backbone and side chain of tyrosine 139, part of the side chain of isoleucine 150 and the side chain of tryptophan 138.

An initial simulation of an isolated molecule, i.e., in the absence of solvent, gave $\Delta\Delta G^{\circ}$ equal to 4 kJ/mole; experimental estimates range between 6 and 9 kJ/mole, the higher value being appropriate for room temperature (B.W. Matthews, J.A Schellman and W. Becktel, unpublished results). As for the isolated α-helix, a part of the calculated value is attributable to the difference in conformational freedom in the denatured state. The observed difference in stability is greater for T4-lysozyme than for the isolated α-helix. The difference is presumably attributable to the difference in the free energy for removing a methyl group from a hydrophobic environment (the protein), and from an aqueous environment (the isolated helix). However, the simulation gives a similar result for the two systems. It is obviously necessary to repeat the work on T4-lysozyme in a solvated system. This calculation is in progress.

A much higher free energy difference was calculated if the conformation of the glycine mutant was artificially kept as that of the wild type protein, i.e., if an empty space was maintained inside the protein. Hence, one concludes that the surrounding side chains gain significantly lower packing energy or higher conformational entropy, or both, by moving into the cavity left by the absent methyl group. It will be interesting to see if the structure will accommodate as easily the introduction of a *larger* side chain, such as valine.

In conclusion, it may be pointed out that the ability to *predict* the effect of protein stability may be applicable to problems of protein engineering. However, for a practical method to emerge, one will also have to learn to deal with difficulties posed by amino acids with long side chains and with charged side chains, by disulfide-bridged proteins and by substitutions causing significant conformation change.

ACKNOWLEDGEMENT. Supported by research and instrumentation grants from the National Science Foundation (DMB 8501037, DMB 8509037). We thank W.F. DeGrado, B.W. Matthews, W. Becktel and J.A. Schellman for providing us with unpublished results.

REFERENCES

Alber, T., Sun, D-P, Wozniak, J.A., Cook, S. & Matthews, B.W., *Nature 330*, 41-46, 1987.

Alber, T., Bell, J.A., Sun, D-P, Nicholson, H., Wozniak, J.A., Cook, S. & Matthews, B.W., *Science 239*, 631-635, 1988.

Anderson, A., Carson, M. & Hermans, J., *Ann. N.Y. Acad. Sci. 482*, 51-59, 1986.

Anderson, A. & Hermans, J., *Proteins, 3*, 262-265, 1988.

Avignon M., Garrigou-Lagrange C. & Bothorel P., *Biopolymers 12*, 1651-1669, 1973.

Bash P.A., Singh U.C., Langridge R. & Kollman P.A., *Science 236*, 564-568, 1987a.

Bash, P.A., Singh, U.C., Brown, S.K., Langridge, R. & Kollman, P.A., *Science 235*, 574-576, 1987b.

Berendsen, H.J.C., Postma, J.P.M. & van Gunsteren, W.F., in *Molecular Dynamics and Protein Structure*, Hermans, J., ed., Polycrystal Book Service, Western Springs, IL, USA, 1985, pp. 43-49.

Beveridge, D.L. & Mezei, M., *ibid.* pp. 53–57.

Brant, D.A. & Flory, P.J., *J. Am. Chem. Soc. 87*, 2788–2791, 1965.

Brant, D.A., Miller, W.G. & Flory, P.J., *J. Mol. Biol. 23*, 47–65, 1967.

Chou, P.Y. & Fasman G.D., *Biochemistry 13*, 211–245, 1974.

Dyson, H.J., Rance, M., Houghten, R.A., Lerner, R.A. & Wright, P.E., *J. Mol. Biol. 201*, 161–200, 1988.

Gray, T.M. & Matthews, B.W., *J. Biol. Chem. 262*, 16858–16864, 1987.

Gunsteren, W.F. van & Berendsen, H.J.C., in *Molecular Dynamics and Protein Structure*, Hermans J., ed., Polycrystal Book Service, Western Springs, IL, USA, 1985, pp. 5–14.

Gunsteren W.F. van & Weiner P.K., eds. *Computer Simulations of Biomolecular Systems*, ESCOM, Leiden, 1989.

Hermans, J. & Puett, D., *Biopolymers 10*, 895–914, 1971.

Hermans, J. & Shankar, S., *Isr. J. Chem. 27*, 225–227, 1987.

Hermans, J, Yun R.H & Anderson, A.G., in preparation, 1989.

Hwang J.-K. & Warshel A., *Biochemistry 26*, 2669–2673, 1987.

Madison V. & Kopple K.D., *J. Am. Chem. Soc. 102*, 4855–4863, 1980.

Marqusee, S. & Baldwin, R.L., *Proc. Natl. Acad. Sci. USA, 84*, 8898–9002, 1987.

Matthews, B.W., Nicholson, H. & Becktel, W.J., *Proc. Natl. Acad. Sci USA 84*, 6663–6667, 1987.

Matsumura, M., Becktel, W.J. & Matthews, B.W., *Nature 334*, 406–410, 1988.

McCammon J.A., Karim O.A., Lybrand T.P. & Wong C.F., *Ann. N.Y. Acad. Sci. 482*, 210–221, 1986.

Rao S.N., Singh U.C., Bash P.A. & Kollman P.A., *Nature 328*, 551–554, 1987.

Rose, G. D., Gierasch, L. M. & Smith, J. A., *Adv. Protein. Chem. 37*, 1–109, 1985.

Scheraga H.A. & Epand, R., *Biochemistry 7*, 2864–2872, 1968.

Straatsma, T.P., Berendsen, H.J.C., & Postma, J.P.M., *J. Chem. Phys. 85*, 6720–6727 , 1986.

113

K.G.Strehlow and R.L. Baldwin, *Biochemistry* 28, 2130-2133, 1989.

Sueki, M., Lee, S., Powers, S.P., Denton, J.B., Konishi , Y. & Scheraga, H.A., *Macromolecules* 17, 148-155, 1984.

Warshel, A., Sussman, F. & King, G. *Biochemistry 25*, 8368-8372. *Phys. Rev.* 136, A405-411, 1986.

Wong C.F. & McCammon J.A., *J. Am. Chem. Soc. 108*, 3830, 1986.

Wong C.F. & McCammon J.A., *Isr. J. Chem. 27*, 217-224, 1987.

Wilmot, C.M. & Thornton, J.M., *J. Mol. Biol. 203*, 221-232, 1988.

Wright, P. E., Dyson, H. J. and Lerner, R. A., *Biochemistry 27*, 7167-7175, 1988.

Zimm, B.H. and J.K. Bragg, *J. Chem. Phys. 31*, 526-535, 1959.

The Use of Molecular Dynamics and Free Energy Perturbation Approaches in Simulating the Properties of Macromolecules and Their Binding to Ligands

PETER A. KOLLMAN

Abstract

In this article we present a review of the use of molecular dynamics in simulating the properties of macromolecules. Specifically, we focus on the use of molecular dynamics in NMR structural refinement, in understanding loop motion and catalytic activity of mutants in triose phosphate isomerase and in the application of free energy perturbation methods in protein design, site-specific mutagenesis and in thermal stability.

Overview

There are two fundamental difficulties in using computer based methods to study complex macromolecules. The first, and most daunting, is the fact that complex molecules have a very large number of degrees of freedom. Currently, one cannot hope to search, let alone accurately evaluate the free energy of all the possible local minima in systems much larger than 100 atoms. Thus, we are a long way from being able to predict protein tertiary structure from amino acid sequence.

Secondly, one wishes to evaluate the energy or free energy of molecular systems to compare with relevant experiments. For molecules with 10 atoms from the first rows of the periodic table, approximate solutions to the Schroedinger equation (*ab initio* calculations) can come close to giving a complete description of the molecule in the gas phase. For complex molecules in solution, one must use a simple, analytical molecular mechanical energy function. Although such functions are incapable of representing quantum mechanical reality, we and others have shown they can effectively represent non-bonded interactions and conformational analysis in molecules. For enzymatic reactions, one can combine quantum mechanical calculations for the parts of the system where bonds are being made/broken with molecular mechanical methods to represent the rest of the system. Such an approach should be very useful in many biophysical simulations.

In this brief review, we focus on a number of current applications of computer simulations from our own laboratory. First, we describe the use of combining simulations with NMR data in structural refinement; secondly, we describe the combined use of quantum mechanics, molecular dynamics and free energy perturbation approaches to study enzyme catalysis and site specific mutagenesis effects on it in triose phosphate isomerase; finally, we describe the application of free energy perturbation methods in drug design, enzyme catalysis and protein thermal stability.

Protein Structural Refinement

One of the most exciting developments in recent years is the use of molecular dynamics to aid in structural refinement. The combination of molecular mechanics and structure factors to refine X-ray derived structures by Brunger and Karplus (Brunger et al., 1987) and van Gunsteren et al. (Fujinaga et al., 1988) has been most impressive and these same two groups have also developed approaches to refine structures derived by the use of NMR/NOE data. (Brunger et al., 1986), (Kaptein et al., 1985)

In the case of the NMR NOE data, there is no general agreement on the "right" way to combine the experimental data with the theoretical calculations to refine the structure or on the criterion for "success". One uses a "pseudo-energy" term of the sort $E = W_1 \sum$ molecular mechanical energies $+W_2 \sum$ experimental fit, where the experimental fit can be the sum of the differences between the calculated and observed NOE's or X-ray structure factors, and W_1 and W_2 are the weights given, respectively, to the two "energy" terms. By making W_2 large enough, one can fit the data well, but perhaps at the cost of a distorted structure. On the other hand, making W_2 too small will lead to a structure that is subject to the inaccuracies of the molecular mechanical energy function as well as the inability to search randomly and find the "correct" local minimum. (Brunger et al., 1987), (Kaptein et al., 1985)

We have been applying combined molecular mechanics/dynamics with NMR NOE data to a study of EETI, a 28 amino acid protein that is a potent trypsin inhibitor. There are fifteen possible disulfide pairings in this molecule and Heitz et al. (Heitz et al., 1989) have suggested which was the correct one, based on distance geometry (DG) calculations. We continued this study (Chiche et al., 1989) by using restrained molecular mechanics and dynamics to try to improve the structure as well as improve the agreement with the NOE data. We found that relative molecular mechanical energies were not a good criterion for which was the "best" structure. The solvation free energy/residue (SFE) was similar for known protein structures. This gave us an additional very useful criteria to evaluate our unknown structures. In addition, we found it critical to carry out our restrained molecular dynamics calculations with an explicit representation of H_2O to create structures with an SFE similar to known structures. This is particularly critical when the number of NOE's/residue is limited, as it was for EETI.

After our structural refinements were completed, a paper appeared in the literature on a related CMTI protein, (Bode et al., 1989) confirming the disulfide pairing predicted earlier and supported by our subsequent studies. We also learned that the structure of EETI complexed with trypsin was being solved. (Mornon, unpublished results) A receipt of the preliminary structure and comparison with our best structure and CMTI suggested that residues 22-25 in the preliminary X-ray structure were incorrectly placed, due to poor resolution of the data in this region. Subsequent refinement of this region using our RMD structure improved the fit to the electron density.

Additionally, we found the structure generated by restrained molecular dynamics in water was far superior to the distance geometry one, and the restrained molecular dynamics structure generated in vacuo in overall fit to the X-ray structure. As noted above, our lab is not a "large player" in the structural refinment area, but the above described study is representative of the power of combined NMR/NOE and X-ray structural refinement of proteins.

Combined Use of Quantum Mechanics, Molecular Mechanics and Molecular Dynamics in the Analysis of Enzyme Action

We have been studying the mechanism of action of triose phosphate isomerase using computer based techniques for some years. Triose phosphate isomerase (TIM) has been characterized as a "perfectly evolved" enzyme because the rate limiting step is product dissociation from the enzyme. The chemical steps have been sped up by $\sim 10^{10}$ due to enzyme catalysis. TIM catalyzes the isomerization of dihydroxyacetone phosphate (DHAP) to glyceraldehyde 3-phosphate, which corresponds to abstraction of a proton from one carbon and its delivery to a neighboring one. It accomplishes this task with such facility through a base, Gln 165 and electrophiles His 95 and Lys 13, which polarize the substrate and facilitate the incipient endiolate anion formation. (Knowles et al., 1977) In fact, our earliest calculations on TIM found that only the Pro-R hydrogen of DHAP was in a region of negative electrostatic potential from the enzyme and the rest of the substrate in a region of positive potential, providing further support for the polarization of the substrate by the enzyme.

Our first study on TIM used a combination of quantum mechanics and molecular mechanics to both study models for the enzymatic reaction and to examine the structure of the enzyme-substrate complex. The X-ray structure used as a starting point for the refinement was at low resolution. It was encouraging that molecular mechanics refinement of the DHAP-native TIM structure remained near the X-ray structure and interesting that a His 95 $\rightarrow$ Gln 95 mutant structural refinement moved the Gln 95 to H bond with Asp 165, thus suggesting a possible reason why a mutant enzyme with His 95 mutated to Gln might be less active than the native His 95 structure. (Alagona et al., 1984)

Subsequently, studies showed that a His 95 $\rightarrow$ Gln 95 mutant was much less active than native TIM, and a Gln 165 $\rightarrow$ Asp 165 mutation was $\sim 10^3$ less active. We carried out quantum mechanical calculations to demonstrate that, if the $-CO_2$ of Glu formed an ideal 2.6Å O ... H-C distance during proton abstraction from DHAP, one needed to postulate only a 0.3Å lengthening of the transition state O ... C distance to rationalize a $\sim 10^3$ or ~ 4 kcal/mole increase in rate constant for proton transfer. (Alagona et al., 1986) Subsequent free energy perturbation calculations examined the alternative possibility that the O ... C distance for the transition state remained 2.6 Å, that the polarizing electrophites His 95 and Lys 13 were less well placed to stabilize the transition state for proton transfer. (Daggett et al., 1989a) It is clear that experiments and calculations cannot rule out either of these possibilities, but the calculations do support the fact that either possibility can easily lead to observed changes in activity by $\sim 10^3$ or so.

We have also used molecular dynamics to study TIM and its mutants. (Brown *et al.*, 1987) There is a critical "loop closing" motion in the catalytic action of TIM and this loop closes only in the presence of DHAP, not the singly charged analog dihydroxyacetone sulfate DHAS. We carried out comparative simulations on TIM without substrate, with DHAP substrate and with DHAS substrate. Even though we used a very low resolution X-ray structure of chicken muscle TIM as a starting model, and did not include H_2O in the simulations, our calculations were consistent with experiments, in that loop closing only occurred with DHAP in the active site. Furthermore, the structure of the loop closed structure was qualitatively consistent with the structure of substrate bound yeast TIM published by Alber *et al.* (1982).

More recently, we have been using molecular dynamics models to try to rationalize the relative catalytic activity of a series of TIM mutants, including revertant structures. For example, whereas mutating Glu 165 $\rightarrow$ Asp 165 reduces k_{cat} by $\sim 10^3$, accompanying this mutation with Ser 96 $\rightarrow$ Pro 96 results in a catalytic rate that is only 20 less than wild type. Mutating Ser 96 $\rightarrow$ Pro 96 alone results in a 10 times lower catalytic activity than wild type. We have been partially successful in rationalizing these relative k_{cat} values by creating structures of models of the transition state for proton abstraction from DHAP by TIM and evaluating the average H-bonding to the key oxygen atoms in the DHAP transition state during 20 psec of molecular dynamics simulations. Those sequences that have strong H-bonding to these oxygens tend to have large k_{cat} values and those with weaker H-bonding tend to have the smaller k_{cat} values. (Daggett *et al.*, 1989b)

In summary, we have found the various computer simulation methodologies -- quantum mechanics, molecular mechanics and molecular dynamics -- useful in analyzing the properties of enzyme catalysis by TIM.

The Use of Free Energy Perturbation Methods in Simulating Inhibitor Binding, Enzyme Catalysis and Protein Stability

[1]

$$
\begin{array}{ccccc}
E+S_1 & \xrightarrow{\Delta G_1} & ES_1 & \xrightarrow{\Delta G_1^{\ddagger}} & E_1TS \\
\Big\downarrow{\Delta G_3} & & \Big\downarrow{\Delta G_4} & & \Big\downarrow{\Delta G_5} \\
E+S_2 & \xrightarrow{\Delta G_2} & ES_2 & \xrightarrow{\Delta G_2^{\ddagger}} & ETS
\end{array}
$$

The thermodynamic cycle above can be used to analyze relative binding of different ligands S_1 and S_2 to an enzyme in either a non-covalent mode ES_i or in a model for the transition state for catalysis ETS_i. The experimental free energies on binding ΔG_1 and ΔG_2 and catalysis $\Delta G_1^{\ddagger}$ and $\Delta G_2^{\ddagger}$ can be determined from the measured K_M and k_{cat} values. These horizontal processes are hard to simulate because they involve (ΔG_i) large changes and movements in water and substrate and $(\Delta G_i^{\ddagger})$ quantum mechanical effects. The vertical processes ΔG_3, ΔG_4 and ΔG_5 are those that we have studied by free energy perturbation methods.

$$\Delta\Delta G_{bind} = \Delta G_2 - \Delta G_1 = \Delta G_4 - \Delta G_3 \qquad [2]$$

$$\Delta\Delta G_{cat} = \Delta G_2^{\ddagger} - \Delta G_1^{\ddagger} = \Delta G_5 - \Delta G_4 \qquad [3]$$

Analogously, we can use cycle [4] to analyze the effect of site specific mutations on the enzyme rather than the ligand on binding and catalysis.

$$[4]$$

Equations [2] and [3] remain valid.

Our first application of this approach to "drug design" involved comparing the free energy of binding of CBZ - Gly^P - X-Leu - Leu (S_1 amidate, X=NH and S_2 ester, X=O) inhibitors of the proteolytic enzyme thermolysin using cycle (1). Bartlett (Bartlett *et al.*, 1987) had found for a variety of amidate (NH) and ester (O) inhibitors of thermolysin that $\Delta\Delta G_{bind} = 4.1 \pm 0.1$ kcal/mole and the crystal structures by B. Matthews *et al.* (Tronrud *et al.*, 1987) showed that the inhibitors bound nearly identically in the active site. The reason for this difference of 4 kcal/mole was attributed to the Ala 113 C=O hydrogen bond with the N-H of the amidate. Our calculations (Bash *et al.*, 1987) on this system led to $\Delta\Delta G_{bind} = 4.2 \pm 0.5$ kcal/mole, with ΔG_4 (binding) = 7 kcal/mole and $\Delta\Delta G$ (solvation) = 3 kcal/mole. We thus suggested that not only were differential solvation important, but that the ester (O) inhibitor was held in place because of the leucines, the CBZ and the phosphate interacting with the Zn^{+2} in what we called "forced repulsion". Thus, to relate the NH vs O free energy difference to the intrinsic strength of a hydrogen bond, was, we felt, misleading.

In order to test this interpretation as well as use the theory in a predictive mode, which is a much more stringent test than reproducing experiment, we agreed with Paul Bartlett to predict the relative binding affinity of phosphinate (X=CH$_2$)

inhibitors. (Merz *et al.*, 1989) Before doing this, we studied the $\Delta\Delta G_{bind}$ (NH vs O) with different simulation protocols and molecular mechanical models and found $\Delta\Delta G$ values ranging from 3.3 to 5.9 kcal/mole. In addition ΔG_3 varied from 0 to 3 kcal/mole depending on models. It is clear that our estimated error of ± 0.5 kcal/mole original reported was optimistic, but also clear that all models were consistent with a large (>500) preference in binding affinity for X=NH vs X=O inhibitors. Two models used in comparing X=NH to X=CH$_2$ and they found, somewhat surprisingly, that both inhibitors were predicted to bind nearly identically to the enzyme ($\Delta\Delta G_{bind}$ = 0.0 to 0.3 kcal/mole favoring NH). A paper describing these results was submitted to JACS in September, 1988 and a copy sent to Bartlett shortly thereafter. His group finished the synthesis of these inhibitors in October and tested a variety of amidate vs phosphinate inhibitors, finding a $\Delta\Delta G_{bind}$ averaging 0.1 kcal/mole, (Bartlett *et al.*, 1989) in amazingly good agreement with the predictions. The calculations had indeed supported the importance of "forced repulsion", since ΔG_4 was only 2 kcal/mole for NH $\rightarrow$ CH$_2$ and 5-7 kcal/mole for NH $\rightarrow$ O, as well as differential solvation, since the X = CH$_2$ inhibitor is 2 kcal/mole easier to desolvate than X=NH. (Merz *et al.*, 1989)

Our first application of cycle [4] was to subtilisin where Rao *et al.* (1987) predicted the effect on binding. $\Delta\Delta G_{bind} = \Delta G_4 - \Delta G_3$ and catalysis, $\Delta\Delta G_{cat} = \Delta G_5 - \Delta G_4$ for the Asn 155 -> Ala mutation in subtilisin. The Genentech group had earlier studied the Asn 155 $\rightarrow$ Thr and Asn 155 $\longrightarrow$ His mutations and had found small $\Delta\Delta G_{bind}$ and large $\Delta\Delta G_{cat}$ values, but in any case the values for the Asn 155 $\rightarrow$ Ala ($\Delta\Delta G_{bind}$ = 0.1 $\pm$ 0.8 (0.4 experimental) and $\Delta\Delta G_{cat}$ = 3.4 $\pm$ 1.0 (3.7 experimental) were in encouraging agreement.

How does one construct the model for E$_i$TS or ETS$_i$ necessary for the calculation of ΔG_5? We have used quantum mechanical calculations on stable tetrahedral adducts of CH$_3$O- and amides to determine molecular mechanical parameters for tetrahedral species and then assumed that these adequately represented the transition state for amide hydrolysis in subtilisin. For this approach to work, it is essential that this tetrahedral intermediate resemble the transition state, that this transition state be similar in E$_1$ and E$_2$ and that the group being mutated not interact too strongly electronically with the atoms in the reactive part of the enzyme. The accuracy with which we calculate $\Delta\Delta G_{cat}$ suggests this is so for this mutation in subtilisin. Hwang and Warshel (1987) have also simulated this and other mutations in subtilisin and have been able to reproduce/predict $\Delta\Delta G_{bind}$ and $\Delta\Delta G_{cat}$ in Asn 155 mutants. Their approach, which uses an empirical valence bond method to simulate bond making/breaking, can in principle take into account electronic structure changes in E$_1$ and E$_2$ directly.

One can use the free energy perturbation approach to analyze sequence dependent stabilities in proteins or nucleic acids. The analysis can be made in terms of the following thermodynamic cycle:

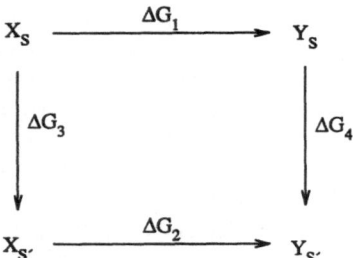

$$X_S \xrightarrow{\Delta G_1} Y_S$$

The relative stability of sequence S in structures X and Y, ΔG_1, can be compared to the relative stability in sequence S', ΔG_2. However, the relative stabilities can be compared to the free energy changes upon mutating S $\rightarrow$ S' in structure X (ΔG_3) and Y (ΔG_4).

For example, we have applied this cycle to study the relative stability toward denaturation of native T4 lyzozyme and its Thr 157 $\rightarrow$ Val mutant. (Dang et al., 1989) In this case, X = native and Y = denatured protein and S = native sequence and S' = mutated sequence Thr 157 $\rightarrow$ Val. Experimentally, one measures ΔG_2

$$\Delta\Delta G_{stab} = \Delta G_2 - \Delta G_1 = \Delta G_4 - \Delta G_3 \qquad [6]$$

and ΔG_1, but in the computer, we can only simulate ΔG_4 and ΔG_3. ΔG_3 is straightforward to simulate, being analogous to ΔG_3 in cycle (4). What does one do to calculate ΔG_4, the mutational free energy for changing Thr 157 $\rightarrow$ Val in the denatured protein, given that one doesn't have a structure for the denatured protein? Our approach was to use a tetrapeptide model of residues 156-158 as a model for the denatured protein, and, doing so, our calculated $\Delta\Delta G_{stab}$ = 1.9 ± 1.1 kcal/mole was in good agreement with the experimental value of 1.6 kcal/mole. Our calculations also suggested that the components contributing to $\Delta\Delta G_{stab}$ were 1.7 kcal/mole van der Waals and only 0.2 electrostatic differential stabilization. This is presumably because the Thr O-H forms just as good H-bonds to water in the denatured state as it does to protein groups in the native protein, but that differential packing in the native structure (OH $\rightarrow$ CH_3) is more favorable for the smaller OH. We confirmed this result by carrying out a model calculation in which only the charges on C_β, O_γ and HO_γ (which give all the H bond stabilization) were mutated from their normal value to zero on O_γ and HO_γ and a small value on C_β to insure charge neutrality. The free energy charge upon this model mutation was within 0.1 kcal/mole of being identical in native and denatured protein. A further "model" mutation zeroed the charges on the backbone NH of Asp 159 in native and mutant proteins to assess the contribution of this group to differential stabilization and the resulting difference (0.6 kcal/mole) suggested that the Asp NH H$\bullet\bullet\bullet$Thr$^{157}O_\gamma$ interaction contributed some to the greater stability of the native (Thr 157) than mutant (Val 157) protein, but did not explain it all. Both of these model calculations emphasize the power of the free energy perturbation method to not only yield numbers, but also useful insight. (Rao et al., 1987)

On the other hand, the theoretical calculations have been less successful in reproducing the observed X-ray structures upon mutation. Mutating Thr → Val results in the movement of the new CH_3 group away from the "H-bonded pocket" of the Thr wild type, but not to the same orientation as observed crystallographically. Mutating Val → Thr does not result in the regeneration of the complex H-bonding network Thr 157 participates in in the wild type X-ray structure, although the Thr 157 O-H hydrogen bonds to water molecules. Other H-bonds, e.g. involving residues 151 and 155 remain intact throughout the simulation. Thus, it is clear that the crystal structures of T4 lyzozyme and its mutants (Alber *et al.*, 1987) will provide an extremely useful testing ground for theoretical simulations.

Summary and Conclusions

In conclusion, we have given examples how computer simulation methods can be used in a complementary way to experimental approaches to understand macromolecular properties.

Acknowledgements

I would like to acknowledge my collaborators, mentioned in the references, the use of the UCSF Computer Graphics Laboratory, NIH-RR-1081, R. Langridge, Director, and research support by DARPA (N00014-86-K-0757), R. Langridge, P.I. and NIH-GM-29072.

References

G. Alagona, K. Ghio, and P. Kollman, *J Amer Chem Soc*, 106:3623, 1984.

G. Alagona, C. Ghio, and P. A. Kollman, *J Mol Biol*, 191:23, 1986.

T. Alber, S. Dao-pin, K. Wilson, J. Wozniak, S. Cook, and B. Matthews, *Nature*, 330:41, 1987. Coordinates deposited in Brookhaven Protein Data Base.

P. A. Bartlett and C. K. Marlowe, *Science*, 235:569, 1987.

P. Bartlett and B. P. Morgan, 1989. Unpublished results.

P. Bash, U. C. Singh, F. Brown, R. Langridge, and P. Kollman, *Science*, 235:574, 1987.

W. Bode, H. J. Greyling, R. Huber, J. Otlewski, and T. Wilusz, *FEBS Lett*, 242:285, 1989.

F. K. Brown and P. A. Kollman, *J Mol Biol*, 198:533, 1987.

A. T. Brunger, G. M. Clore, A. M. Gronenborn, and M. Karplus, *Proc Natl Acad Sci USA*, 83:3801-3805, 1986.

A. Brunger, J. Kuriyan, and M. Karplus, *Science*, 235:458, 1987.

L. Chiche, C. Gaboriaud, A. Hertz, J. P. Mornon, B. Castro, and P. Kollman, "Use of Restrained Molecular Dynamics in Water to Determine 3D Structure: Prediction of the 3D structure of Trypsin Inhibitor EETI-II," *Proteins*, 1989. Submitted .

V. Daggett, F. K. Brown, and P. A. Kollman, "Free Energy Component Analysis:

A Study of the Glu 165 → Asp 165 mutation of Triose Phosphate Isomerase,'' *J Amer Chem Soc*, 1989a. In press.

V. Daggett and P. Kollman, "Molecular Dynamics Simulations of Mutants and Revertants of Triose Phosphate Isomerase: Rationalization of Relative Catalytic Potencies," 1989b. Manuscript in preparation.

L. Dang, K. Merz Jr, and P. Kollman, "Free Energy Calculations of Protein Stability: The Thr 157 → Val 157 Mutation to T4 Lyzozyme," *J Amer Chem Soc*, 1989. In press.

M. Fujinaga, P. Gros, and W. van Gunsteren, *J Applied Crystallography*, 1988. In press .

A. Heitz, L. Chiche, D. Le-Nguyen, and B. Castro, *Biochemistry*, 28:2392, 1989.

J. K. Hwang and A. Warshel, *Biochemistry*, 26:2669, 1987.

R. Kaptein, E. Zuiderweg, R. Scheek, R. Boelens, and W. Gunsteren, *J Mol Biol*, 182:179, 1985.

J. Knowles and W. Albery, *Acc Chem Res*, 10:105, 1977.

P. Kollman, S. Rao, F. Brown, V. Daggett, G. Seibel, and U. C. Singh, "Free Energy Perturbation Methods Can Give Exciting Insights Into the Effect of Site-Specific Mutants on Both Binding and Catalysis: Applications to Subtilisin, Trypsin and Triose Phosphate Isomerase and the Description of a Free Energy Component Analysis," *Protein Structure, Folding and Design 2*, p. 215-225, A. Liss, Inc., 1987.

K. Merz and P. Kollman, "Free Energy Perturbation Simulations of the Inhibition of Thermolysin: Predictions of the Free Energy of Binding of a New Inhibitor," *J Amer Chem Soc*, 1989. in press .

J. P. Mornon, unpublished results.

S. Rao, P. Bash, U. C. Singh, and P. Kollman, *Nature*, 328:551-554, 1987.

D. E. Tronrud, H. M. Holden, and B. W. Matthews, *Science*, 235:571, 1987.

Molecular Recognition of DNA Minor Groove Binding Drugs

ANDREW H.-J. WANG and MAI-KUN TENG

ABSTRACT

The molecular structure of several complexes between three minor groove binding drugs (netropsin, distamycin and Hoechst 33258) and several related DNA dodecamers has been solved and refined by single crystal X-ray diffraction analysis. In these complexes, the drug molecules bind to the central AT segment (4-6 base pairs long) in the narrow minor groove of the dodecamer B-DNA double helix. The stabilizing forces between the drugs and DNAs are provided by a combination of ionic, van der Waals and hydrogen bonding interactions. Bifurcated hydrogen bonds are found between the drugs and DNA dodecamers with A_nT_n sequence in which the AT base pairs have large propeller twists. In contrast, no bifurcated hydrogen bond is found in the drug-d(CGCGATATCGCG) complex due to the unique dispositions of the hydrogen bond acceptors (N3 of adenine and O2 of thymine) of the AT base pairs on the floor of the DNA minor groove. In the latter case, two of the four AT base pairs in the ATAT stretch have low propeller twist angles, even though the DNA has a narrow minor groove. In the netropsin-d(CGCGATATCGCG) structure, the drug is found to occupy in two orientations equally well, suggesting a disordered model. This is consistent with the results from solution studies (chemical footprinting and NMR) of the dynamic binding interactions between minor groove binding drugs and DNA. A simplified model is presented for the complete binding process of the drug to DNA based on these structural results.

INTRODUCTION

DNA double helix is the target molecule of many antitumor drugs. In fact, DNA molecules may be considered as the ultimate receptors for these drugs. There has been a great deal of interest in understanding the molecular basis of those antitumor drugs, e.g., their binding affinity and specificity (Wang, 1987). This is particularly important in that the correlation of the biological activities of the drugs with the manner in which the drug molecules bind the DNA double helix may provide new ways to improve their chemotherapeutic properties.

An essential part of the comprehensive understanding of the interactions between these antitumor drugs and DNA (the receptor molecule) is the conformational range which the DNA molecules are capable of adopting. Over the years, there has been significant advances in our understanding of the way in which DNA can adopt different conformations depending on the its nucleotide sequence and many other extrinsic factors. For example, the left-handed Z-DNA double helix is favored by alternating C-G sequence (Wang, et al., 1979). The interconversion between Z-DNA and the right-handed B-DNA is influenced by metal ions, ionic strength, supercoiling and Z-DNA binding proteins as reviewed elsewhere (Rich, et al. 1984; Wells and Harvey, 1987). Similarly, sequences with a string of guanines may have a propensity to adopt an A-DNA conformation (Wang, et al., 1982; Thomas & Wang, 1988). Another conformational state of DNA, namely bent DNA, has received a great deal of attention due to its potential role in gene regulation and nucleosome phasing (Widom, 1985). Bent DNA is apparently sequence directed and is strongly favored by oligo-$(dA)_n$ [n=4-6] stretches when they are appropriately and repeatedly spaced along DNA (Diekmann, 1987). However, some subtle idiosyncrasies exist. For example, DNA made of repeats of a (GAAAATTTTC) motif exhibits a strong bent DNA tendency, but not that built of (GTTTTAAAAT) (Burkhoff and Tullius, 1988). The reason for this may be related to the dinucleotide steps in the center, TpA vs ApT.

More recently, it also became clear that many DNA binding molecules can have profound influence on DNA conformation. For example, quinoxaline antibiotics (triostine A and echinomycin), which are bis-intercalating molecules, have a strong affinity to tetranucleotide sequences with a CpG step in the middle of the sequences. The specificity for CpG sequence is provided by the hydrogen bonds between the two alanine residues in the octadepsipeptide backbone of the quinoxaline antibiotics and the NH_2 amino group of guanine in the minor groove of the highly distorted right-handed double helix (Wang, et al., 1984; 1986). When triostin A is bound to the DNA molecule, it can induce the base pairs adjacent to the intercalator quinoxaline ring into an alternative Hoogsteen geometry. These results underlies the importance of the full appreciation of DNA polymorphism.

A number of natural and synthetic compounds are known to bind to DNA double helix in a non-intercalative manner. Several well-known compounds of this type, such as the antitumor antibiotics netropsin and distamycin, and the synthetic Hoechst 33258, are shown in Figure 1. The molecular basis of the interactions between them and DNA have recently been under intensive study by a number of methods including chemical

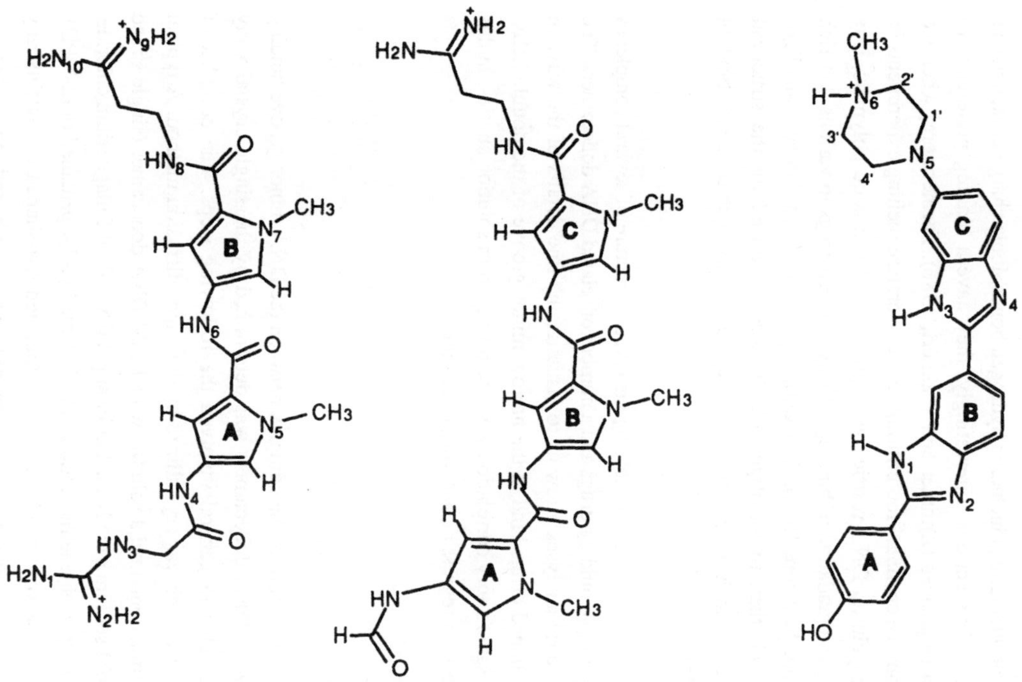

Netropsin   Distamycin A   Hoechst 33258

Figure 1. Chemical formula of netropsin, distamycin A and Hoechst 33258. Note that the molecules are asymmetric. The aromatic rings in the molecules are labelled A, B and C. All molecules have similar crescent shape which allows the molecules to fit the contour surface of the B-DNA minor groove.

footprinting experiments (Taylor, et al., 1984; Dervan, 1986; Ward, et al., 1988), NMR (Patel, 1978; Klevit, et al., 1986; Leupin, et al., 1986; Pelton and Wemmer, 1988) and single crystal X-ray diffraction studies (Kopka, et al., 1985; Coll, et al., 1987; 1989; Teng, et al., 1988; Carrondo, et al., 1989). The picture emerging from these studies is that this class of compounds binds in the narrow minor groove of the B-DNA double helix using a combination of interactions including hydrogen bonds, ionic charge attractions, as well as van der Waals interactions. Interestingly, they have a binding preference to stretches of AT-rich sequences (Zimmer and Wahnert, 1986). Furthermore, when we survey the results from the chemical footprinting and temperature melting experiments of netropsin and distamycin on DNA restriction fragments and synthetic polymers, it is evident that there exists a gradation of binding affinity of these drugs to various A-T rich sequences (Zimmer and Wahnert, 1986). For example, netropsin binds better to an oligo-d(A).oligo-d(T) stretch than to an oligo-(dA-dT) stretch. Based on the structural information available so far, it is not immediately apparent why such sequence binding microheterogeneity exists.

We have approached these problems by solving the crystal structure of several complexes between those minor groove binding drugs and a number of related DNA dodecamers. The results of those structural analyses allow us to visualize the fine details of the ways in which the crescent-shaped drugs bind in the narrow minor groove of the double helix. They also provide a satisfactory correlation with the results from solution studies. In this paper, we summarize the progress of our work in this area.

EXPERIMENTALS

The three dimensional structure of the complex between the DNA minor groove binding compounds and several DNA dodecamers was determined by the single crystal x-ray diffraction analysis. This in general involved the following steps. The complex of netropsin and the dodecamer was crystallized by the vapor diffusion method. A typical crystallization condition contained a solution with 1 mM DNA dodecamer (single strand concentration), 40 mM sodium cacodylate buffer at pH 6.5, 7.0 mM magnesium chloride, 2.0 mM drug, 4.5 mM spermine tetrachloride and 5% 2-methyl-2,4-pentanediol (2-MPD), equilibrated against a reservoir of 50% 2-MPD. Elongated rod-shaped crystals usually appeared in one to three weeks. Diffraction data were collected on automated diffractometers.

It can be seen from Table1 that the overall unit cell dimensions, and consequently the diffraction patterns, are very similar among those closely related drug-dodecamer complexes, suggesting a very similar crystal lattice for all the complexes. A model of the

Table 1. Crystallographic data of DNA dodecamers and
their complexs with minor groove binding compounds

DNA	Unit Cell Dimension (Å)			R-factor	Resolu-tion(Å)	Reference
	a	b	c			
CGCGAATTCGCG	24.87	40.39	66.20	0.18	1.9	Drew, et al. (1981)
CGCGAATTCGCG +Hoechst 33258	25.04	40.33	65.85	0.14	2.2	Pjura, et al. (1987)
CGCGAATTCGCG +Hoechst 33258	25.23	40.58	66.08	0.15	2.2	Teng, et al. (1988)
CGCGAATT[Br]CGCG +Netropsin	24.27	39.62	63.57	0.21	2.2	Kopka, et al. (1985)
CGCGATATCGCG +Netropsin	25.48	41.26	66.88	0.20	2.4	Coll, et al. (1989)
CGCGATATCGCG +Hoechst 33258	25.59	40.56	67.10	0.13	2.3	Carrondo, et al. (1989)
CGCAAATTTGCG +Distamycin A	25.20	41.07	64.65	0.20	2.2	Coll, et al. (1987)
CGCAAAAAAGCG CGCTTTTTTGCG	25.40	40.70	65.80	0.20	2.5	Nelson, et al. (1987)

dodecamer was generated and refined using the Konnert-Hendrickson refinement
procedure (Hendrickson and Konnert, 1979) to a reasonably low R-factor (~30%). At this
stage, solvent molecules located from the difference Fourier map were gradually included
in the refinement. However, we refrained from adding any solvent molecules in the minor
groove region of the double helix to avoid any interference in the interpretation of the
difference Fourier map there. After the inclusion of some water molecules in addition to
the dodecamer duplex, the R-factor was reduced to low 20's%. At this stage, the
difference Fourier map clearly showed distinct residual electron density in the minor
groove at the AT region. The drug molecule was then placed in the residual density in the
minor groove of the DNA double helix using computer graphic system with the program
FRODO (Jones, 1978) and included in the model. The refinement was continued with the
DNA, drug and solvent molecules until it converged with a final R-factor in the range of
14-19%. This procedure results in a reliable medium-resolution (1.5-2.5 Å) structure as
illustrated in Figure 2 where the drug molecule is seen to fit the final "omit" Fourier map
very well.

RESULTS AND DISCUSSION

Overall Structure of the Drug-DNA Complexes

The overall structure of the drug-DNA complex is represented by the distamycin-
d(CGCAAATTTGCG) structure shown with a van der Waals diagram looking into the

128

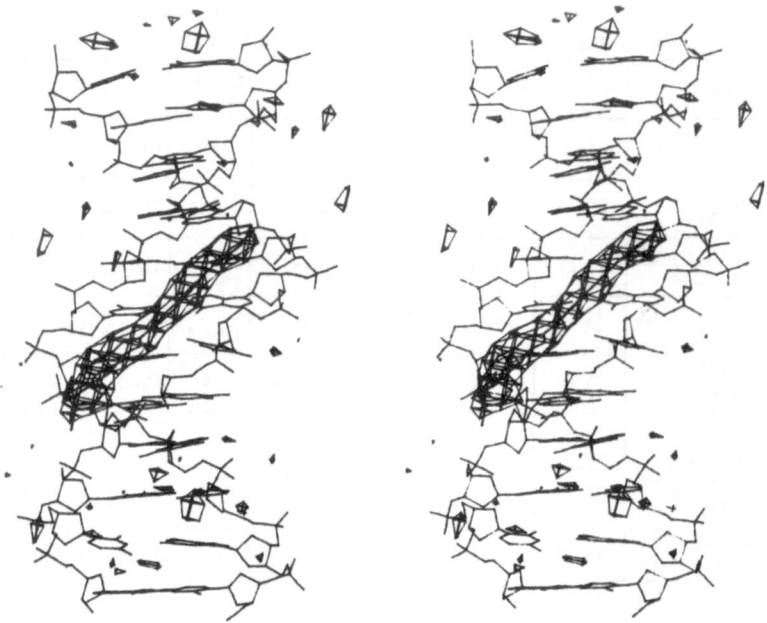

Figure 2. Final "omit" difference Fourier electron density map associated with the Hoechst 33258 molecule in the crystal of the Hoechst-d(CGCGATATCGCG) complex displayed by the FRODO program. on an IRIS 4D/70GT graphic system. Phases were calculated based on all the atoms in the structure except the Hoechst molecule.

minor groove of the right-handed B-DNA double helix in Figure 3. It can be seen that the elongated distamycin molecule fits tightly in the narrow minor groove at the central AT region covering slightly over five base pairs. The drug molecule is sandwiched by the side wall of the minor groove which is made of the two anti-parallel sugar-phosphate backbones. Many atoms from DNA have close van der Waals contacts to both faces of the flat distamycin molecule. The drug molecules have significant twists due to the large dihedral angles between the pyrrole-amide units which are necessary for the distamycin molecule to follow the contour surface of the B-DNA minor groove. The dihedral angles in these binding drugs are listed in Table 2.

See color insert:

Figure 3. Van der Waals diagram of the distamycin-d(CGCAAATTTGCG) complex looking into the minor groove along the direction of the molecular pseudo diad axis. The elongated flat distamycin fits snugly in the narrow minor groove with many van der Waals contacts to the side walls of the groove.

Table 2. Dihedral Angle between Planes in DNA Binding Drugs

Drugs	A · B [*]	B · C [*]	Reference
CGCGAATTCGCG-Hoechst 33258	9.0°	30.1°	Teng, et al. (1988)
CGCGAATTCGCG-Hoechst 33258	0.5°	37.1°	Pjura, et al. (1987)
CGCGATATCGCG-Hoechst 33258	18.5°	13.0°	Carrondo, et al. (1989)
CGCGATATCGCG-Netropsin	22.6°	——	Coll, et al. (1987)
CGCGAATT [Br] CGCG-Netropsin	24.8°	——	Kopka, et al. (1985)
CGCAAATTTGCG-Distamycin A	9.9°	16.9°	Coll, et al. (1987)

[*] A, B and C are the aromatic rings shown in Figure 1. Rings A and B include the amide group preceding the pyrrole ring.

A more detailed depiction of the structure is illustrated by the stereoscopic skeletal view (Figure 4A) which has the intermolecular hydrogen bonds shown as thin lines. This is a view with the same orientation as in the van der Waals diagram of Figure 3. There are several hydrogen bonds between the drug and the nucleophilic oxygen and nitrogen atoms of DNA at the floor of the minor groove. The distamycin amide NH groups form an array of bifurcated hydrogen bonds to DNA. The large number of bifurcated hydrogen bonds in the complex are possible due to the fact that the AT base pairs in the A_3T_3 sequence adopt a highly propeller twisted conformation such that the disposition of N3 and O2 atoms of A_3T_3 sequence at the floor of the minor groove are optimized for such hydrogen bonding interactions.

Another example is the complex of netropsin with d(CGCGATATCGCG) shown in Figure 4B. As in the structure of distamycin complex, the crescent-shaped netropsin drug hugs closely to the DNA double helix with the pyrrole HC5 and HC11 hydrogens approaching the HC2 hydrogens of adenine bases. For example, the HC2 atom of pyrrole A is 2.6 Å away from the HC2 atom of adenine A7 and the HC11 atom of pyrrole B is 2.6 Å from the HC2 atom of adenine A19 (all hydrogen atom positions were calculated at their theoretical positions). These close van der Waals contacts imply that there would be serious clashes between these drugs and guanine bases in the minor groove, if AT is replced vy GC base pairs. As suggested previously, the N2 amino group of a guanine base in the minor groove presents a major hindrance for the entry of the drugs into the minor groove, thereby providing the discrimination for a more favored binding toward AT base pairs (Kopka, et al., 1985). The netropsin in this complex completely covers the ATAT segment with its linear guanidinium and amidinium ends touching the two GC outer base pairs (G4-C21 and C9-G16). Again this agrees well with the previous observations made by the footprinting experiments (Taylor, et al. 1984; Ward, et al., 1988).

A

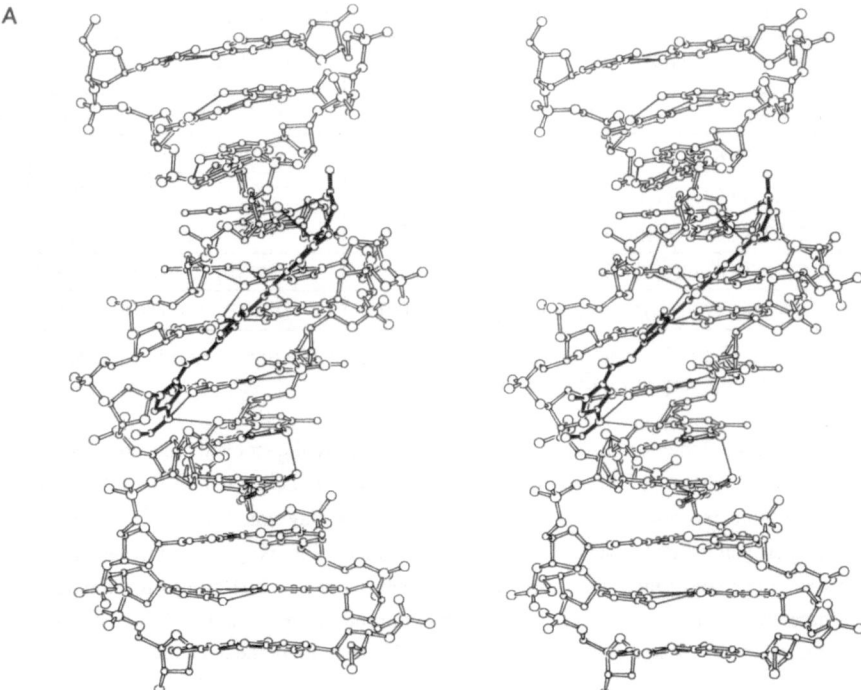

Figure 4. Stereoscopic skeletal diagrams of (A) the distamycin-d(CGCAAATTTGCG) complex and (B) the netropsin-d(CGCGATATCGCG) complex with DNA in open bonds

Netropsin and distamycin

Both netropsin and distamycin are N-methylpyrrole-containing antitumor antibiotics which have elongated flat crescent-shaped geometry. Nature has designed those molecules ingeniously by optimizing their functional groups for the binding in the B-DNA minor groove. For example, the nitrogen atom on the pyrrole ring is blocked by a methyl group such that this nitrogen atom will not be involved in the undesirable hydrogen bonding interactions. Instead the proper hydrogen bonding donors are provided by the NH groups of the amide linkages.

The detailed hydrogen bonding interactions of netropsin and distamycin with three different DNA dodecamers are summarized in Figure 5. It can be seen that the drugs netropsin and distamycin form more hydrogen bonds with AATT and AAATTT sequences respectively than the netropsin with the alternating ATAT sequence. In the netropsin-d(CGCGAATTCGCG) and distamycin-d(CGCAAATTTGCG) complexes, the NH

B

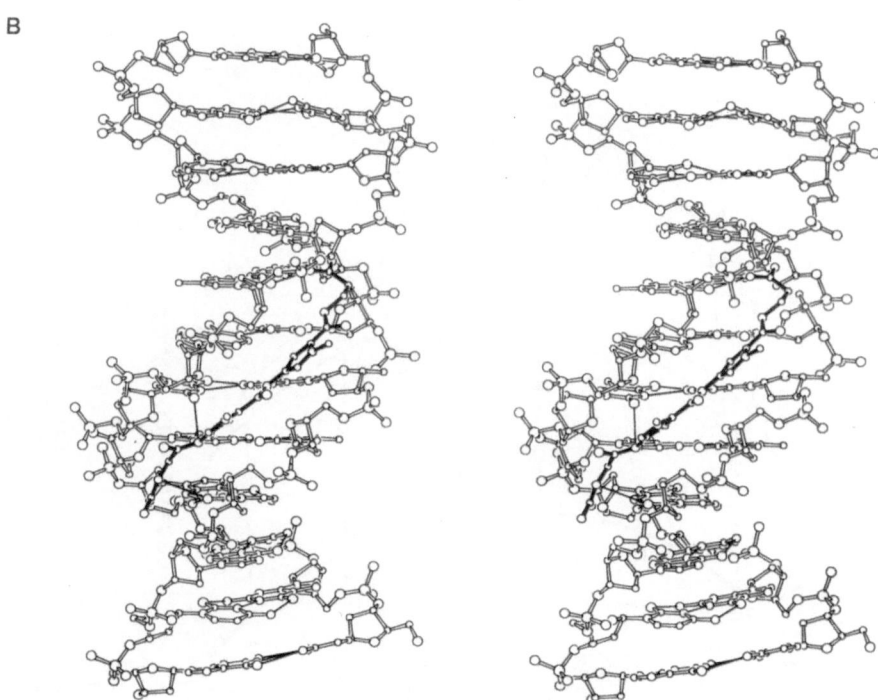

and the drugs in filled bonds. The elongated distamycin and netropsin molecules lie in the narrow minor groove of the double helix with hydrogen bonds to bases of DNA.

groups from the peptide bonds form a series of bifurcated (three-centered) to the N3 of adenine and O2 of thymine bases. There is an interesting pattern in the complex of distamycin-d(CGCAAATTTGCG). Notice that two consecutive NH groups (N5 and N7) can form bifurcated hydrogen bonds to three adjacent AT base pairs. However, the third NH group (N3) is no longer in the hydrogen bonding range to the AT base pair. This is presumably due to the natural curvature of this type of pyrroleamide-containing compounds does not fit the contour surface of the B-DNA double helix perfectly. Interestingly, the next (fourth) amide NH group returns to the proper position to continue this bifurcated hydrogen bonds down the helix. Consequently, distamycin, which has three pyrrole rings and four NH groups, can bind to a DNA segment of five base pairs long. This observation agrees well with the results from the chemical footprinting experiments using long synthetic distamycin-like oligomers containing more than 3 pyrrole amide units. In those experiments, it was found that the number of base pairs that the synthetic distamycin analog molecule can protect is one larger than the number of the amide NH groups. This pattern has been dubbed the so-called (n+1) rule (Dervan, 1986).

132

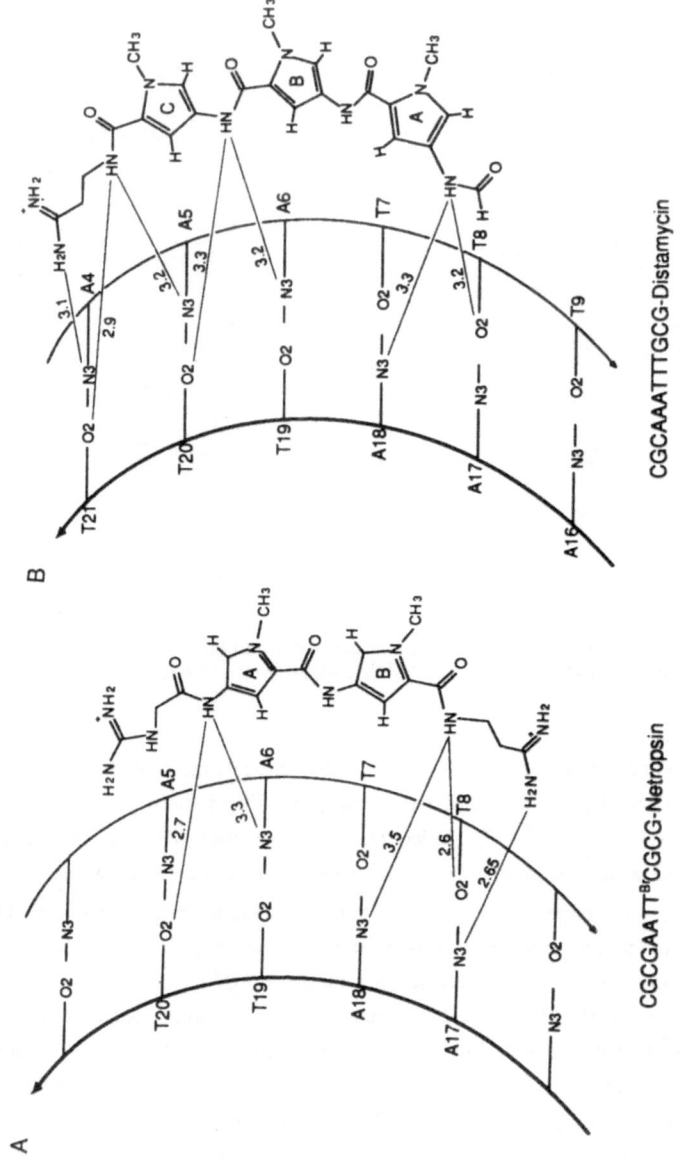

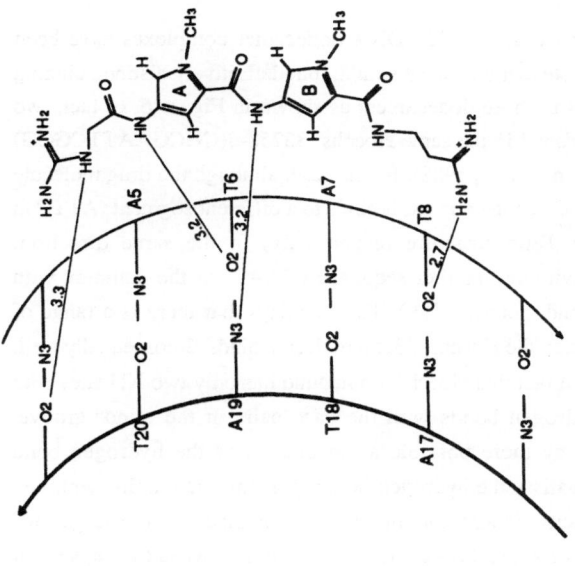

CGCGATATCGCG-Netropsin

Figure 5. Comparative illustration of different complexes of DNA with netropsin /distamycin. (A) Netropsin-d(CGCGAATTCGCG) complex. There are five hydrogen bonds between the netropsin and DNA, of which four are bifurcated hydrogen bonds from the amide NH groups to base O2 and N3 atoms (Kopka et al., 1985). (B) Distamycin-d(CGCAAATTTGCG) complex. Three sets of bifurcated hydrogen bonds are found (Coll et al., 1987). (C) Netropsin-d(CGCGATATCGCG) complex. This is one the two possible orientations of the complex in the crystal lattice. Only single hydrogen bonds are found due to the unique dispositions of N3 and O2 atoms in the alternating ATAT sequence (Coll et al., 1989).

Our results here also suggest that we may replace the one of every three amide linkages with an ester bond which may result in a GC instead of AT sequence specificity. This is due to the fact that the replacement of the NH group in the amide bond by the oxygen atom in the ester linker could eliminate the potential van der Waals clashes between the NH of the drug and the NH2 of the guanines. Most of the attentions thus far in the design of the so-called "lexitropsin" have been directed toward replacing the pyrrole ring with other five- or six-membered rings such as furan or pyridine. A combination of the modifications both in the aromatic rings and in the linkers of the molecules may provide a better binding specificity toward more general nucleotide sequences.

Synthetic compound: Hoechst 33258

Three independent analysis of Hoechst 33258-DNA dodecamer complexes have been carried out. It is interesting to note that they have similar, but definitively distinct, binding mode toward the AT segments in those dodecamers as shown in Figure 6. In fact, two different binding modes were found in the same Hoechst 33258-d(CGCGAATTCGCG) complex (Pjura, et al., 1987; Teng, et al., 1988). In this case, although the drug molecule covers same length (4 nucleotides) of the duplex, it binds to a different segment (ATTC in Pjura structure vs AATT in Teng structure respectively) in the same direction. Furthermore, the drug binds with yet another sequence (GAAT) in the complex with d(CGCGATATCGCG) (Carrondo, et al., 1989). This suggests that there is a range of tetranucleotide sequences to which the Hoechst 33258 molecule binds almost equally well. This may be associated with the fact that Hoechst compound has only two NH sites with which it can use to form hydrogen bonds with the base pairs in the minor groove. Therefore there would be many more possible arrangements of the hydrogen bond acceptors in DNA which can satisfy the hydrogen bonding geometric requirement, i.e., less stringent sequence specificity. In addition, the molecule is made of four rings (two benzimidazoles, one phenol and one piperazine). The connections between the rings result in a more straight molecule which does not follow the groove surface nearly as well in comparison with the natural antibiotics netropsin and distamycin.

The primary contribution of the binding affinity of Hoechst compound to DNA is probably not from the hydrogen bonding interactions, but is likely from the many van der Waals interactions between the sugar-phosphate backbone of DNA with the aromatic ring of the drug. This is due to the very narrow minor groove in the AT region as will be described later. This is clearly illustrated in Figure 7 where some atoms in the sugar rings can be seen to have close van der Waals contacts with the Hoechst compound (Teng, et al., 1988). For example, the O4' of T20 residue is only 3.2 Å away from the plane of the netropsin pyrrole ring, close to the sum of the van der Waals radius of oxygen (1.4 Å) and the half thickness of an aromatic pyrrole ring. As suggested previously, this sandwiching of the planar drug by the sugar-phosphate backbones from the sidewalls of minor groove of B-DNA double helix strongly stabilizes the complex and may be considered as a pseudo-intercalation (Teng, et al., 1988).

Influence of drugs on DNA conformation

<u>Minor groove width</u>

The overall conformation of the DNA dodecamer duplexes with closely related sequences listed in Table 1 is quite similar to each other, though they differ in their fine details. The averaged torsion angles of the drug-DNA complexes are very similar to those of the canonical d(CGCGAATTCGCG) double helix, which has the conformation typical of the right-handed B-DNA. However, they all possess a prominently characteristic feature of a very distinct narrow minor groove at the central AT region. This is easily seen by comparing the minor groove width of all the known crystal structures of dodecamer duplexes diagrammatically shown in Figure 8. A consistent trend is obvious in these curves where they all reach the lowest values (~4 Å) near the 8/21 residue, except for the AATT dodecamer without a bound drug.

This is associated with the fact that all dodecamers crystallized in this lattice possess quite similar helical parameters. For example when we inspect the roll angles of six dodecamers crystallized in this lattice, while they vary somewhat from molecule to molecule, it is clear that the overall trend is the same in all structures. Foremost are the two steps with large roll angles, namely the C3-G4 step with an average of +8 degrees and G10-C11 step with an average of -9 degrees. It is likely that this more a consequence of the lattice packing interactions, rather than the intrinsic properties of a particular dinucleotide step.

However, we should emphasize that the narrow minor groove is likely to be an intrinsic feature associated with AT-rich sequences. We have recently solved the crystal structure of another dodecamer duplex made of three separate strands [d(CGCGAAAACGCG)+d(CGCGTT)+d(TTCGCG)] (unpublished results). The resulting duplex has a missing phosphate in one strand, therefore it can serve as a good model for a nicked DNA. The molecule crystallizes in a different crystal lattice, yet it still has the same distinctive narrow minor groove. Another example was seen in the complex of the phage 434 repressor protein with a 22-mer containing O_R1 operator sequence, which is very rich in AT base pairs (Aggarwal, et al., 1988). The DNA molecule in this complex has a minor groove width ranging from ~9 Å in the central AT region to ~14 Å near the end of the helix. There are many highly propeller-twisted base pairs in this helix, not unlike what we have seen in the distamycin-d(CGCAAATTTGCG) complex (Coll, et al., 1987) as will be described below.

136

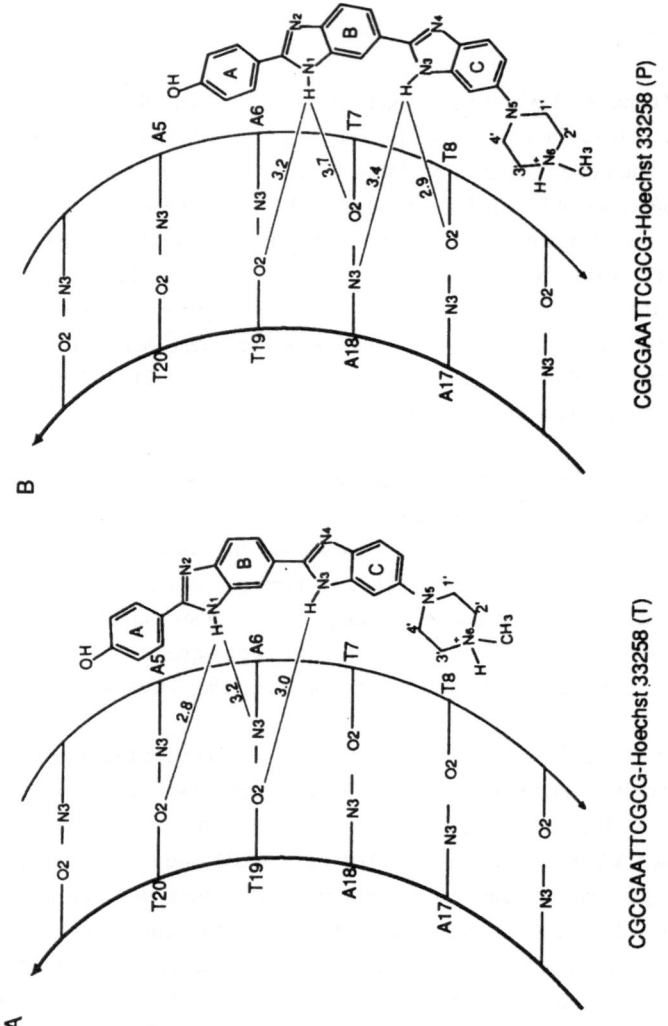

B CGCGAATTCGCG-Hoechst 33258 (P)

A CGCGAATTCGCG-Hoechst 33258 (T)

137

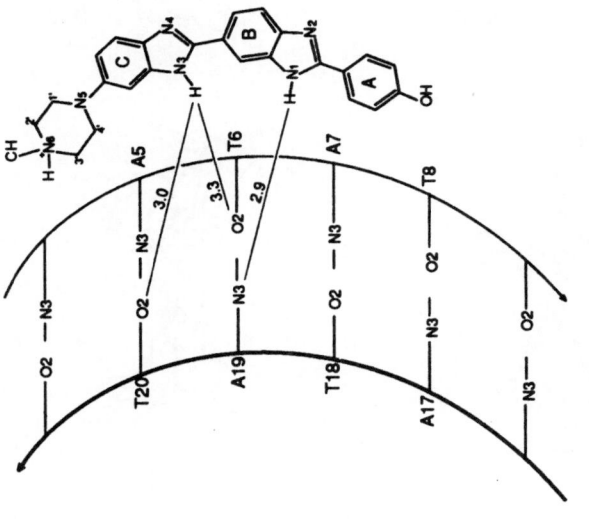

CGCGATATCGCG-Hoechst 33258

Figure 6. Comparisons of the interactions between Hoechst 33258 and DNA dodecamers. (A) and (B) show respectively the interaction between the Hoechst dye and the same DNA dodecamer d(CGCGAATTCGCG) (Teng, et al., 1988; Pjura, et al., 1987). Note that the dye molecule binds in the same orientation, but it covers different base pairs in the dodecamer duplex. (C) Hoechst 33258-d(CGCGATATCGCG) complex. In this structure, the Hoechst dye points in the opposite direction comparing to the complexes of (A) and (B). Notice there are very few hydrogen bonds between Hoechst dye and the DNA base pairs in all three complexes.

138

A

B

C

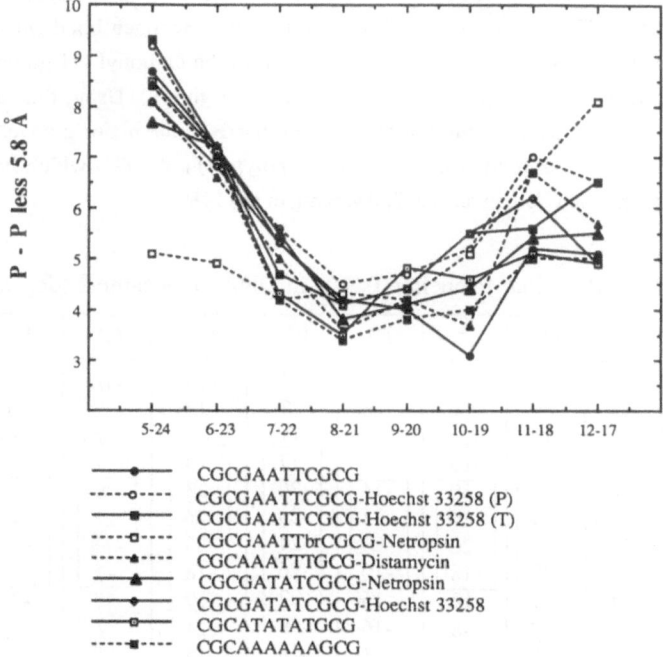

Minor Groove Width in DNA Dodecamer

Figure 8. A composite plot of the phosphate to phosphate distances across the minor groove in various dodecamers. The closest distance is near P8/P21 and P9/P20 in all cases.

Propeller twist angles

One of the interesting features found in some of these drug-DNA complexes is that there is a large propeller twist (>20°) in the A_nT_n segments of the DNA dodecamers. This is clearly illustrated in Table 3 which compares the propeller twist angle of the AT vs GC base pairs in these related dodecamers. It can be seen that the A_nT_n sequences have high

Figure 7. (A) A stereo view near the Hoechst dye binding site showing in detail the van der Waals interactions between the drug and DNA. The drug is viewed edged on and the large twist between the two benzimidazole rings is evident. The O4' atoms of the sugar are pointing toward the flat aromatic surface of the Hoechst dye. The base pairs have been removed for clarity. (B) A view 90° from (A). This view clearly shows the manner in which the planar drug is sandwiched between the two sugar-phosphate backbones, analogous to the intercalation binding of intercalator in DNA double helix. This stabilization is possible only when the minor groove is very narrow. (C) A portion of the cylindrical plot of the complex indicates all the close van der Waals contacts between DNA sugar O4' atoms and the Hoechst dye.

propeller twist (ω) angles (molecules *A-F* in Table 3). In some structures, the high ω values of certain AT steps have resulted in some very interesting intramolecular hydrogen bonds. In those AT steps, a number of novel bifurcated hydrogen bonds in the major groove between the N6 amino group of an adenine to the carbonyl O4 groups of two adjacent thymines of the opposite strand are formed (Figure 9). Using this nucleotide structural motif incorporating bifurcated hydrogen bonds in the major groove, we were able to construct a modified B-DNA model of poly(dA)·poly(dT), which has a very narrow minor groove (Coll, et al., 1987; Aymami, et al., 1989).

Table 3. Base Pair Propeller Twists in DNA Dodecamers (degrees)

Base pair	A[a]	B[a]	C[a]	D[a]	E[a]	F[b]	G[a]	H[a]	I[b]
1-24	-17	-36	-18	-5	-4	-19	-10	6	
2-23	-13	-19	-21	-6	-2	-12	0	-10	
3-22	-4	-14	-22	1	-11	-8	-7	-18	
4-21	-17	-19	-19	3	-17	-15	2	9	
5-20	-27	-21	-30	-14	-20	-23	8	-3	
6-19	-27	-23	-26	-22	-25	-26	-23	-26	
7-18	-24	-25	-20	-20	-24	-23	-13	-10	
8-17	-28	-30	-18	-10	-26	-18	-4	1	
9-16	-25	-21	-18	-16	-22	-19	-17	-32	
10-15	-9	-22	-8	-16	-8	-11	2	-5	
11-14	-27	-40	-26	-21	-18	-15	-28	-15	
12-13	-5	7	5	-6	-15	-6	-8	-9	
Average of all base pairs	-18.6	-21.9	-18.4	-11.0	-16.0	-16.3	-8.2	-9.3	-13.9
Average of AT base pairs	**-26.5**	**-24.8**	**-23.5**	**-16.5**	**-22.3**	**-20.7**	**-8.0**	**-9.5**	**-16.1**
Average of GC base pairs	**-14.6**	**-20.5**	**-15.9**	**-8.3**	**-9.7**	**-11.8**	**-8.3**	**-9.3**	**-11.6**

A. CGCGAATTCGCG   *Drew, et al.,* (1981)
B. CGCGAATT[Br]CGCG-Netropsin   *Kopka, et al.,* (1985)
C. CGCGAATTCGCG-Hoechst 33258   *Pjura, et al.,* (1987)
D. CGCGAATTCGCG-Hoechst 33258   *Teng, et al.,* (1988)
E. CGCAAATTTGCG-Distamycin   *Coll, et al.,* (1987)
F. CGCAAAAAAGCG   *Nelson, et al.,* (1987)
G. CGCGATATCGCG-Netropsin   *Coll, et al.,* (1989)
H. CGCGATATCGCG-Hoechst 33258   *Carrondo, et al.,* (1989)
I. CGCATATATGCG   *Yoon, et al.,* (1988)

[a] Value calculated using program **CURVE.** *Lavery & Sklenar,* (1989)
[b] Published value with sign changed to conform with the new convention. [EMBO J. *8,* 1 (1989)]

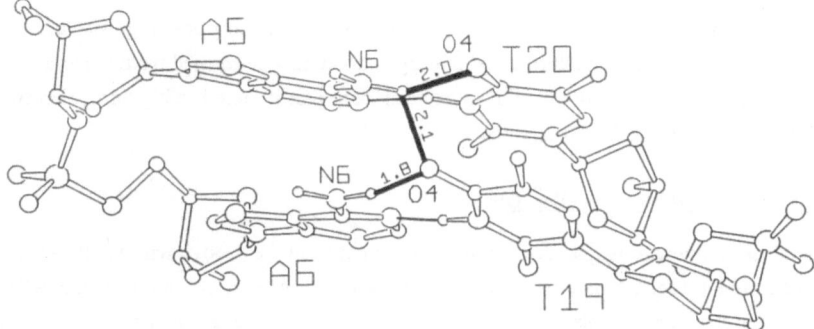

Figure 9. Skeletal diagram showing the bifurcated (three-centered) hydrogen bonds associated with the large propeller twist of A5-T20 and A6-T19 steps. This type of bifurcated hydrogen bond involving the N6 of adenine and the O4 of two thymines in the major groove in an oligo(dA) stretch may be important in the bent DNA structure as discussed in the text.

In contrast, the mean propeller twist angle for the alternating AT base pairs (molecules G-I) is only 11°, a low value compared to those found in other dodecamers and about the same as that found in the GC stretch. It has been suggested previously that the narrow minor groove is associated with the high propeller twist of AT base pairs which can be stabilized by the spine of hydration or drug molecules (Wing, et al., 1980; Kopka, et al., 1985). The structures G-I listed in Table 3 have low averaged propeller twist angle in the AT region, which has lead us to ponder what is the principal factor responsible for the narrow minor groove.

This may be explained as the following. A continuous high propeller twist is unlikely to occur in an alternating AT stretch because of the clash of the N6 atoms of consecutive adenines from opposite strands in the major groove of DNA. This is evident in the stacking diagram of the consecutive AT steps. We shall use the structure G in Table 3 as an example. Here it is interesting to note in this DNA molecule that there is very little base overlap (both inter- and intra-strand) in the middle T6pA7 step as might be expected for the pyrimidine(3'-5')purine step in B-DNA. It is somewhat surprising to see these two base pairs both have relatively high propeller twist (16°) resulting in a very close A7C2 to A19C2 distance (2.7 Å) in the minor groove.

The two outer base pairs (A5-T20 and T8-A17) have very small propeller twist angles (8° and -4° respectively), which are associated with the close adenine-N6 to adenine-N6 distances. The A5N6 to A19N6 distance is 3.3 Å in the A5pT6 step, while the equivalent distance A7N6 to A17N6 distance is 3.1Å in the A7pT8 step. From this data, it can be understood why an alternating AT sequence tends to have low propeller twist angles. On the other hand, in an oligo(dA) stretch, the N6-N6 repulsion does not exist between adjacent base pairs; instead a bifurcated hydrogen bond may be formed from the amino N6 of an adenine to the O4 of thymine in an adjacent base pair if there is a high propeller twist in the stretch (Coll, et al., 1987; Nelson, et al., 1987).

Minor groove width vs propeller twist

This raises an interesting question on the relation between the groove width and base pair propeller twist angle. It is clear that the narrow minor groove observed in the ATAT dodecamer structures (molecules G-I in Table 3) is not the result of the high propeller twist, unlike those seen in the A_nT_n structures (molecules A-F in Table 3). We are not certain what dictates the groove width of a helix. It seems that the characteristics of the minor groove of B-DNA helix is dependent on the base composition and sequence. The groove may be more easily distorted (e.g. compressed) for AT-rich sequences. It is conceivable that many conformational states, including those associated with normal and narrow minor grooves, are at similar energies and have only small barriers between them. Therefore, the crystal packing forces may allow only those conformers with a narrow minor groove of many related dodecamer molecules to pack in this orthorhombic lattice (Table 1). Only the molecules having a narrow groove can be accommodated in this lattice, because there are certain close inter-molecular contacts between two phosphate groups (P2/P7 of 7.4 Å and P10/P18 of 8.1 Å) from two symmetry-related dodecamers. It is worth noting that both P7 and P18 are in the central AT stretch. Those close P-P contacts may restrict the DNA conformation to a very defined range in order to pack them together. In other words, molecules with a wide minor groove would have severe crowding between those phosphate groups in the crystal.

It is interesting to note that the narrow minor groove may be stabilized by groove binding drugs like netropsin. It should be pointed out that the DNA dodecamer d(CGCGATATCGCG) without any drug could not be crystallized in this orthorhombic lattice, supporting the notion that perhaps the ATAT tract has a wider minor groove without a bound drug. On the other hand, dodecamers with $(dA)_n$ tracts in the middle crystallize readily in this orthorhombic lattice with and without groove binding drugs suggesting that the narrow minor groove already exists in solution stabilized by the high propeller twist of the A-T base pairs. It is also interesting to note that another related dodecamer sequence d(CGCTTTAAAGCG) did not crystallize with or without any drug. This may be related to the unique TpA step in those sequences which has been suggested to be disruptive to DNA bending (Diekmann, 1987; Burkhoff and Tullius, 1988).

Sequence micro-specificity of drugs

It is now well known that netropsin and distamycin bind preferentially to the minor groove of the B-DNA helix at the AT-rich regions (Zimmer and Wahnert, 1986). From the high resolution chemical footprinting experiments, it can be further shown that those drugs bind better to an oligo(dA) stretch than to an alternating AT stretch (Taylor, et al., 1984; Dervan, 1986). The molecular basis for this difference may be understood from the dispositions of the N3 of adenine and the O2 of thymine on the floor of the minor groove for various AT sequences. In the structure of the complex of distamycin and d(CGCAAATTTGCG) (Coll, et al., 1987), the distances between the successive N3 and O2 (or O2 to O2) of two neighboring AT base pairs range from 3.4 to 4.4 Å. Those separations are ideally suited for a three-centered (bifurcated) hydrogen bond by placing a hydrogen bond donor (usually an NH group from the amide linkage) halfway between the two acceptor atoms near the bottom of the groove. Water molecules can also fulfill this role, thereby connecting a run of A's through a spine of hydration network. The result of this is that several hydrogen bonds can be formed between the drug and DNA. For example, there are five hydrogen bonds in the the netropsin-d(CGCGAATTCGCG) complex (Kopka, et al., 1985), while there are seven in the distamycin-d(CGCAAATTTGCG) complex (Coll, et al., 1987).

In the alternating AT sequence, the separations between two adjacent N3 and O2 atoms became highly uneven. In the current structure, the O2 to O2 distance for two pairs of thymines are 4.8 and 4.7 Å respectively for T6---T20 and T8---T18, whereas the N3---N3 distance is much shorter (3.3 Å) for A7---A19 bases. It becomes difficult for the hydrogen bond donor NH group to form a bifurcated hydrogen bond for either the O2---O2 pair or the N3---N3 pair. In the former case, the NH group would have to get quite close to both acceptor atoms in order to attain the necessary distance (2.7-3.1 Å) for the hydrogen bond. This would lead to close van der Waals clashes from other parts of the drug molecule. A converse effect would be in place for the latter situation. There the shorter N3---N3 distance would cause the drug molecule to fall away from the DNA, thereby destabilizing the binding of drug molecule. The net effect is that the netropsin adjusts itself to form single, rather than three centered bifurcated, hydrogen bonds between NH of netropsin and O2 (and N3) of DNA. However, this adjustment perturbs the precise registration of the successive NH groups, which are optimally spaced in the slightly curved netropsin molecule for bifurcated interactions, to their corresponding acceptors at the floor of the minor groove. As a consequence, fewer hydrogen bonds can be formed between netropsin (and other minor groove binding compounds) and alternating A-T DNA sequences (see

144

Figures 5 and 6). This provides a satisfactory explanation at the molecular level for the different AT binding affinity of netropsin to DNA observed by the chemical footprinting experiments. Figure 10 summarizes schematically the ways in which these binding drugs interact with the AT stretch of the DNA dodecamers. It clearly illustrates the multitude of their DNA binding modes with respect to the sequences.

Binding Dynamics

Many studies have suggested that the binding of these drugs and DNA is a dynamic process. The results from the crystal structure of netropsin-d(CGCGATATCGCG) showed a two fold disordered netropsin in the crystal lattice. This observed disorder

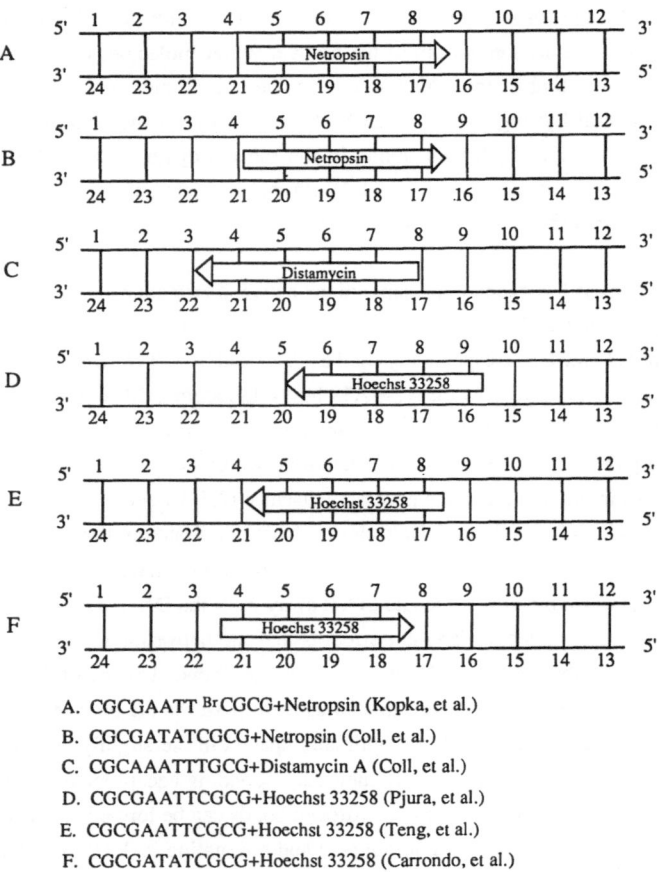

A. CGCGAATT ^{Br}CGCG+Netropsin (Kopka, et al.)

B. CGCGATATCGCG+Netropsin (Coll, et al.)

C. CGCAAATTTGCG+Distamycin A (Coll, et al.)

D. CGCGAATTCGCG+Hoechst 33258 (Pjura, et al.)

E. CGCGAATTCGCG+Hoechst 33258 (Teng, et al.)

F. CGCGATATCGCG+Hoechst 33258 (Carrondo, et al.)

Figure 10. A summary schematic diagram showing the relative position of the binding site of different drugs in various complexes.

reinforces the experimental interpretation of chemical affinity cleaving studies on restriction DNA fragments by synthetic distamycin-like molecules which have an EDTA-Fe(II) cleaving function attached at one end. From the cutting pattern on the high resolution gel, it was suggested that distamycin binds to DNA in two orientations with an asymmetric extent of cleavage at either ends of the binding site (Taylor, et al., 1984). This is also in complete accord with several NMR studies in which the minor binding compounds are shown to interact with DNA double helix in a "flip-flop" fashion on the NMR time scale (Patel, 1979; Klevit, et al., 1986; Leupin, et al., 1986; Pelton & Wemmer, 1988). More recently, we have also shown that a symmetric synthetic pyrrole containing molecule, P1-F4S-P1, binds to the dodecamer d(CGCA3T3GCG) in a similar manner as do netropsin and distamycin by NMR studies. We observed strong nuclear Overhauser effect (NOE) signals between the pyrrole proton and the CH2 proton of *two* adjacent adenines (unpublished result). This may imply that on the NMR time scale the synthetic drug molecule occupies more than one single site in the minor groove. Therefore one can envision a dynamic process existing in solution where the asymmetric netropsin hops on and off the self-complementary double helix to create a non-symmetric complex which can deposit itself onto the crystal surface of the nucleating crystals in the $P2_12_12_1$ lattice in either one of the two orientations. This would generate a statistically two-fold disordered structure. This type of two-fold disordered structure is not uncommon in oligonucleotide crystals as shown by the nanomer d(GGATGGGAG).d(CTCCCATCC) in the A-DNA conformation (McCall, et al., 1986) and the B-DNA dodecamer d(CGCAAAAATGCG)·d(CGCATTTTTGCG) duplex (DiGabriele, et al., 1989).

It is interesting to note that four other complexes, netropsin-d(CGCGAATTCGCG) (Kopka, et al., 1985), distamycin-d(CGCAAATTTGCG) (Coll, et al., 1987), Hoechst 33258-d(CGCGAATTCGCG) (Pjura, et al., 1987; Teng, et al., 1988) and Hoechst 33258-d(CGCGATATCGCG) (Carrondo, et al., 1989) crystallize in the isomorphous lattice in only one unique orientation. In the case of the distamycin complex, this may be due to the hydrogen bond between the guanidinium end of distamycin and the O3' of G24 which may predispose a slight preference for this particular alignment. Similar small but significant interactions may be in operation for the other three complexes. Those structures also support the notion that either orientations in the crystal are energetically very similar. If no specific lattice interactions could be established to bias the stabilization of a particular orientation, then a disordered structure could easily be formed. Figure 11 shows the subtle distortions of the DNA double helix induced by the binding of the drug molecule, as clearly expressed by the path of the helix axis. It can be seen that the DNA molecules adjust their local conformation differently according to the sequence and the type of drug in the complexes.

146

CGCGAATTCGCG   CGCGAATTCGCG+Hoechst 33258

CGCGATATCGCG+Netropsin   CGCAAATTTGCG+Distamycin

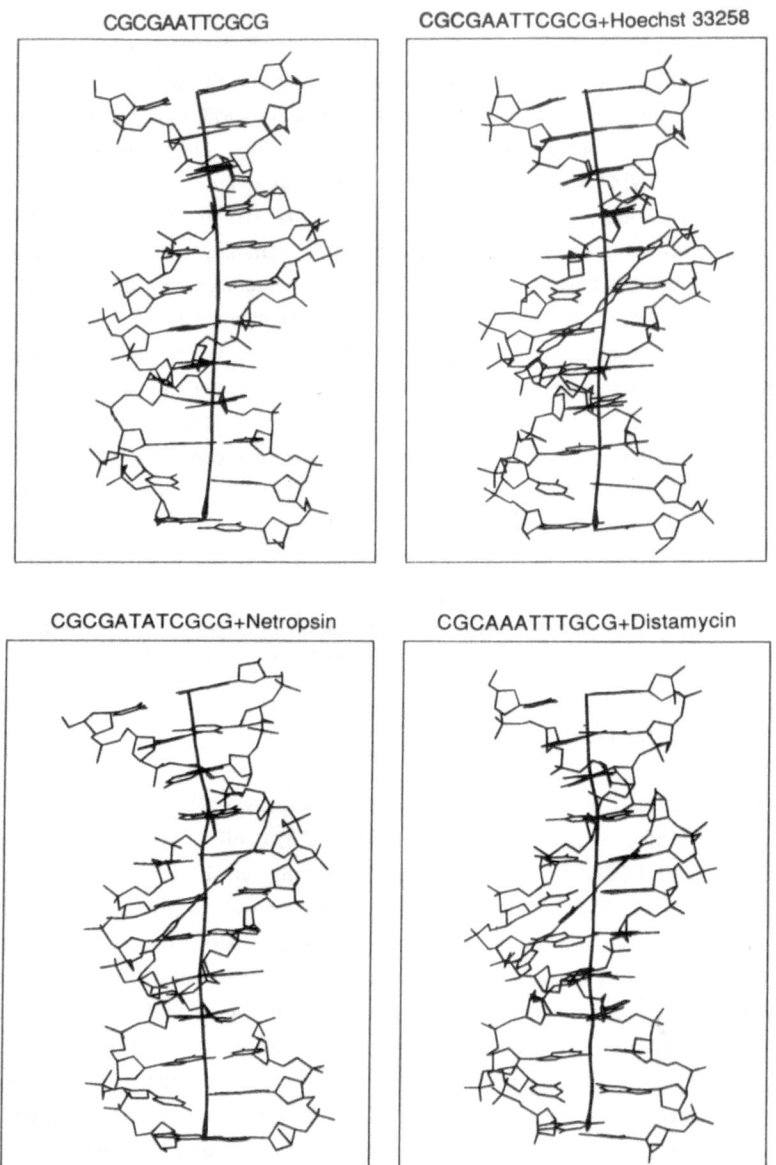

Figure 11. Comparative diagrams showing the helix axis of different dodecamers. It can be seen that the helix axis is almost straight in all structures. However, the drug-DNA complexes seem to have more local perturbations in their helix axis directions.

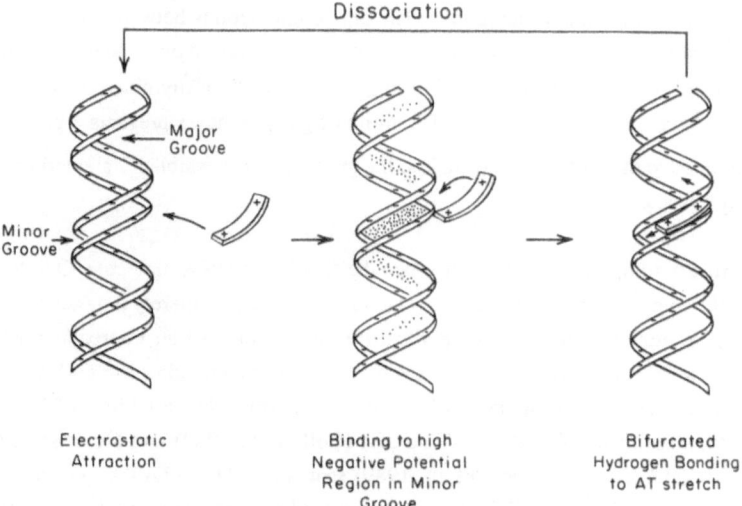

Dissociation

Major Groove

Minor Groove

| Electrostatic Attraction | Binding to high Negative Potential Region in Minor Groove | Bifurcated Hydrogen Bonding to AT stretch |

Figure 12. A highly schematic diagram for the binding of the minor groove binding drugs with DNA double helix. Three major steps are shown. The drug is attracted to the DNA molecule non-specifically by the electrostatic forces. The drug molecule may fall into the minor groove where it has a more negative electrostatic potential. Subsequently, the flat elongated drug may slide up or down in the minor groove along the DNA molecule. GC-rich regions would be unfavorable for binding due to the van der Waals clashes between the pyrrole CH and the guanine NH$_2$ groups.

CONCLUSIONS

The comparison between the structures of the complexes between the drugs and DNA oligonucleotides allows us to visualize the fine details on the manner in which a minor groove binding drug interacts with a B-DNA double helix. Finally, we may summarize the processes by which the drugs bind to the DNA double helix in the minor groove at the AT regions, as shown schematically in Figure 12. There are four major factors which are responsible for the AT binding preference for netropsin, distamycin and related compounds. First, all of them are positively charged molecules, thereby enhancing the initial non-specific attraction between the drug and DNA. In addition, it has been suggested that the deepest negative charge potential of a DNA double helix resides near the bottom of the minor groove at the AT-region (Zarkezewska, et al., 1983). The drug molecule may be driven into the vicinity of the AT region by such potential. The third factor is the AT sequence may be associated with a higher tendency to adopt a narrow minor groove which would provide many van der Waals stabilizing interactions with the sugar-phosphate side walls of the groove. The dipole interactions between the O4' atoms with the aromatic pyrrole π-electron clouds immobilize the drug in place so that the amide NH group can form hydrogen bonds to N3 of adenine and O2 of thymine on the floor of the minor groove. Finally, the amino NH2 group of guanine bases presents a prominent steric hindrance to the entry of the drug into the groove, hence establishing a definite AT binding preference.

This dynamic binding process also illustrates the fluidity of DNA molecule. Our results show that the role of a drug like netropsin extends beyond being merely a passive ligand, which may be attracted to the binding sites. Instead, much like an intercalator which extends the helix length by 3.4 Å, netropsin is capable of inducing the collapse of a "normal" minor groove into a "narrow" groove, suggesting the flexibility of DNA as shown in the netropsin-d(CGCGATATCGCG) (Coll et al., 1989) and Hoechst 33258-d(CGCGATATCGCG) (Carrondo, et al., 1989) structures. The ATAT sequence in the dodecamer d(CGCGATATCGCG) alone may have a low intrinsic propeller twist angle which may disfavor the formation of a narrow minor groove. The binding of netropsin to the ATAT segment will allow the sugar-phosphate backbones to come in close van der Waals contacts with netropsin to create a narrow groove, while still maintaining the low propeller twist in the AT base pairs. The readjustment of the backbone would cost slightly extra energy which may account partly for the low affinity of netropsin to alternating A-T sequence, in comparison to that for an oligo(A) sequence. This kind of small but significant conformational rearrangement undoubtly can also be achieved by proteins interacting with DNA.

Finally, the TpA step may destabilize a run of high propeller twist base pairs in the AT-rich tract. Its presence would tend to disrupt the propagation of a base pair with high propeller twist which would facilitate narrowing of the minor groove. One prediction is that the alternating poly(dA-dT).poly(dA-dT) double helix would resemble the mixed sequence B-DNA structure with a normal minor groove width. Thus it would be less likely to accommodate molecules like netropsin due to the TpA steps (50%) in the polymer.

ACKNOWLEDGEMENTS

This work was supported by a grant from NSF (DMB 8612286) to A.H.-J.W. Helpful discussions from our colleagues, Drs. J. Aymami, M. Coll, C. A. Frederick, A. Rich, are highly appreciated. We also thank Drs. J. H. van Boom and G. A. van der Marel for their continuous supports.

REFERENCES

Aggarwal, A. K., Rodgers, D. W., Drottar, M., Ptashne, M. and Harrison, S. C. (1988) *Science 242*, 899-907.

Aymami, J., Coll, M., Frederick, C. A., Wang, a. H.-J. and Rich, A. (1989) *Nucleic Acids Res.* (in press).

Burkhoff, A. M. and Tullius, T. D. (1988) *Nature 331*, 455-457.

Carrondo, M., Coll, M., Aymami, J., Wang, A. H.-J., van der Marel, G. A., van Boom, J. H. and Rich, A.(1989) *Biochemistry* (in press).

Coll, M., Frederick, C. A., Wang, A. H.-J. and Rich, A. (1987) *Proc. Nat. Acad. Sci. 84*, 8385-8389.

Coll, M., Aymami, J., van der Marel, G. A., van Boom, J. H., Rich, A. and Wang, A. H.-J. (1989) *Biochemistry 28*, 310-320.

Dervan, P. B. (1986) *Science 232*, 464-471.

DiGabriele, A. D., Sanderson, M. and Steitz, T. A. (1989) *Proc. Natl. Acad. Sci. 86*, 1816-1820.

Diekmann, S. (1987) *Nucleic Acids Mol. Biol. 1*, 138-156.

Drew, H. R. and Dickerson, R. E. (1981) *J. Mol. Biol. 149*, 761-786.

Hendrickson, W. A. and Konnert, J. (1979) in *Biomolecular Structure, Conformation, Function and Evolution*, ed. Srinvasan, R. (Pergamon, Oxford), pp. 43-57.

Jones, T. A. (1978) *J. Appl. Crystallogr. 11*, 268-272.

Klevit, R. E., Wemmer, D. E. and Reid, B. R. (1986) *Biochemistry 25*, 3296-3303.

Kopka, M. L., Yoon, C., Goodsell, D., Pjura, P. and Dickerson, R. E. (1985) *Proc. Nat. Acad. Sci. 82*, 1376-1380.

Lavery, R. and Sklenar, H. (1989) *J. Biomol. Struct. Dyn. 6*, 655-667.

Leupin, W., Chazin, W. J., Hyberts, S., Denny, W. A. and Wuthrich, K. (1986) *Biochemistry 25*, 5902-5908.

McCall, M., Brown, T., Hunter, W. N. and Kennard, O. (1986) *Nature(London) 322*, 661-664.

Nelson, H. C. M., Finch, J., Luisi, B. F. and Klug, A. (1987) *Nature(London) 330*, 221-226.

Patel, D. J. (1978) *Eur. J. Biochem. 99*, 369-379.

Pelton, J. G. and Wemmer, D. E. (1988) *Biochemistry 27*, 8088-8096.

Rich, A., Nordheim, A. and Wang, A. H.-J. (1984) *Ann. Rev. Biochem. 53*, 791-846.

Pjura, P., Grzeskowiak, K. and Dickerson, R. E. (1987) *J. Mol. Biol. 197*, 257-271.

Taylor, J. S., Schultz, P. G. and Dervan, P. B. (1984) *Tetrahedron 40*, 457-465.

Teng, M.-K., Frederick, C. A., Usmann, N. and Wang, A. H.-J. (1988) *Nucleic Acids Res. 16*, 2671-2690.

Thomas, Jr., G. J. and Wang, A. H.-J. (1988) *Nucleic Acids Mol. Biol.* 2, 1-30.

Wang, A. H.-J. (1987) *Nucleic Acids Mol. Biol.* 1, 53-69.

Wang, A. H.-J., Fujii, S, van Boom, J. H. and Rich, A. (1982) *Proc. Nat. Acad. Sci.* 79, 3968-3972.

Wang, A. H.-J., Quigley, G. J., Kolpak, F.J., Crawford, J. L., Van Boom, J. H., van der Marel, G. A. and Rich, A. (1979) *Nature* (London) 282, 680-686.

Wang, A. H.-J., Ughetto, G., Quigley, G. J., Hakoshima, T. van der Marel, G. A., van Boom, J. H. and Rich, A. (1984) *Science* (Washington, D.C.) 225, 1115-1121.

Wang, A. H.-J., Ughetto, G., Quigley, G. J., and Rich, A. (1986) *J. Biomol. Struct. Dyn.* 4, 319-342.

Ward, B., Rehfuss, R., Goodisman, J. and Dabrowiak, J. C. (1988) *Biochemistry*, 27, 1198-1204.

Wells, R. D. and Harvey, S. C. (eds) (1987) *Unusual DNA Structures*, Springer-Verlag New York Inc, New York.

Widom, J. (1985) *BioEssays* 2, 11-14.

Wing, R., Drew, H., Takano, T., Broka, C., Tanaka, S., Itakura, K. and Dickerson, R. E. (1980) *Nature* (London) 287, 755-758.

Yoon, C., Prive, G. G., Goodsell, D. S. and Dickerson, R. E. (1988) *Proc. Nat. Acad. Sci.* 85, 6332-6336.

Zakrzewska, K., Lavery, R. and Pullman, B. (1983) *Nucleic Acids Res.* 11, 8825-8839.

Zimmer, C. and Wahnert, U. (1986) *Prog. Biophys. Mol. Biol.* 47, 31-112.

Color Insert

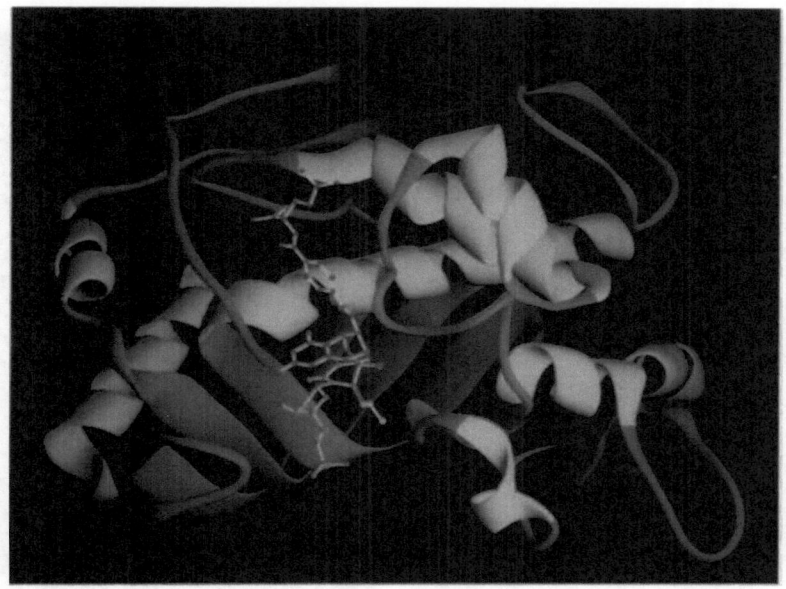

Figure 4 from "Inhibitor Binding to Thymidylate Synthase Is Mediated by Different Structural Determinants than Those That Promote Tight Binding to Dihydrofolate Reductase," by David A. Matthews, Cheryl A. Janson, and Ward W. Smith, p. 5.

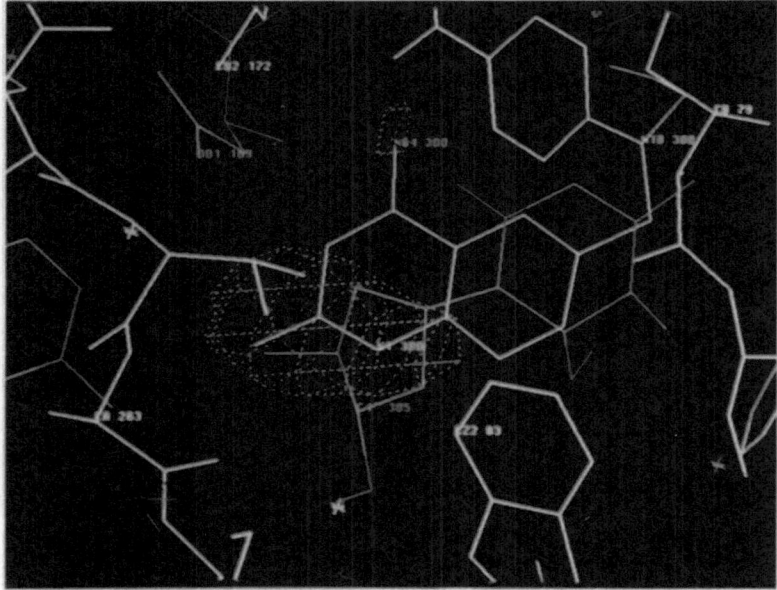

Figure 8 from "Inhibitor Binding to Thymidylate Synthase Is Mediated by Different Structural Determinants than Those That Promote Tight Binding to Dihydrofolate Reductase," by David A. Matthews, Cheryl A. Janson, and Ward W. Smith, p. 7.

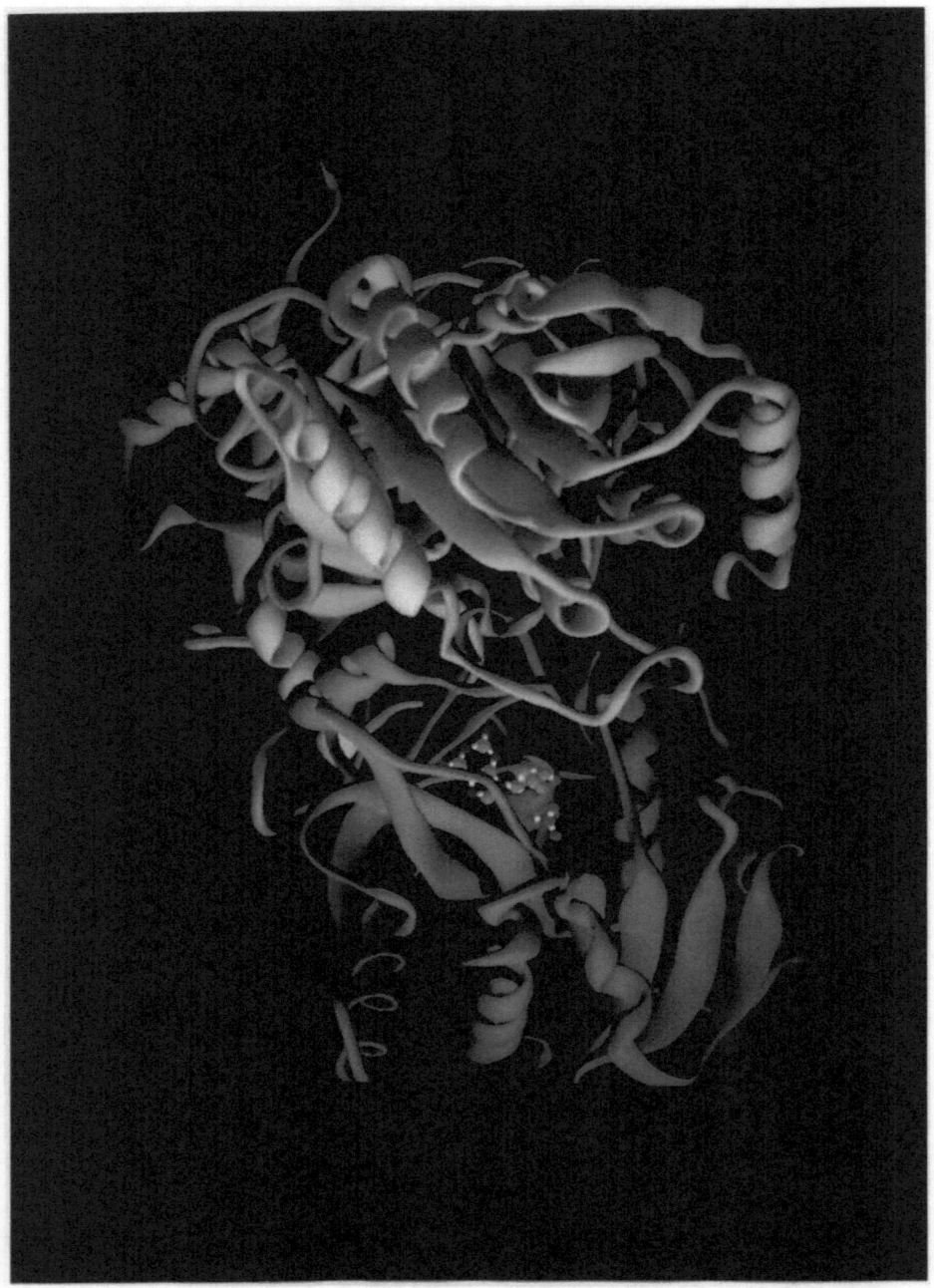

Figure 4 from "Design of Purine Nucleoside Phosphorylase Inhibitors Using X-Ray Crystallography," by Steven E. Ealick, Y.S. Babu, S.V.L. Narayana, William J. Cook, and Charles E. Bugg, p. 47.

Figure 5 from "Design of Purine Nucleoside Phosphorylase Inhibitors Using X-Ray Crystallography," by Steven E. Ealick, Y.S. Babu, S.V.L. Narayana, William J. Cook, and Charles E. Bugg, p. 47.

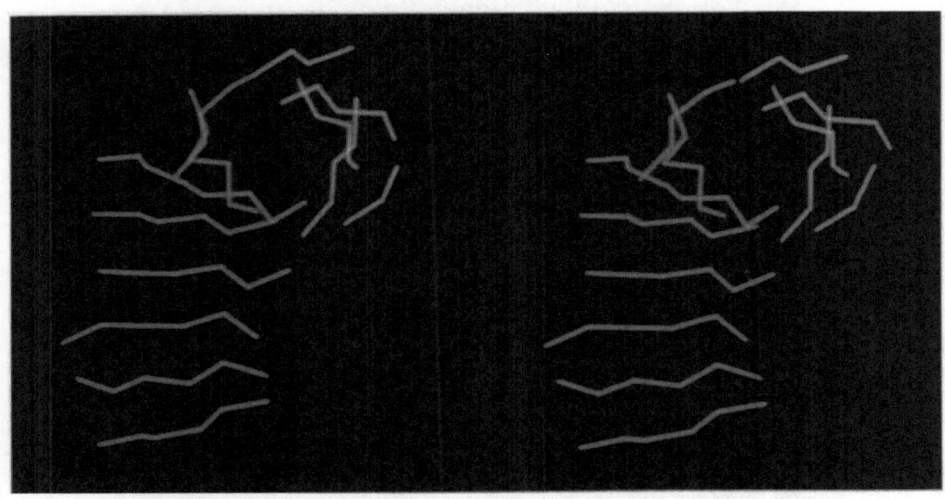

Figure 6 from "Design of Purine Nucleoside Phosphorylase Inhibitors Using X-Ray Crystallography," by Steven E. Ealick, Y.S. Babu, S.V.L. Narayana, William J. Cook, and Charles E. Bugg, p. 47.

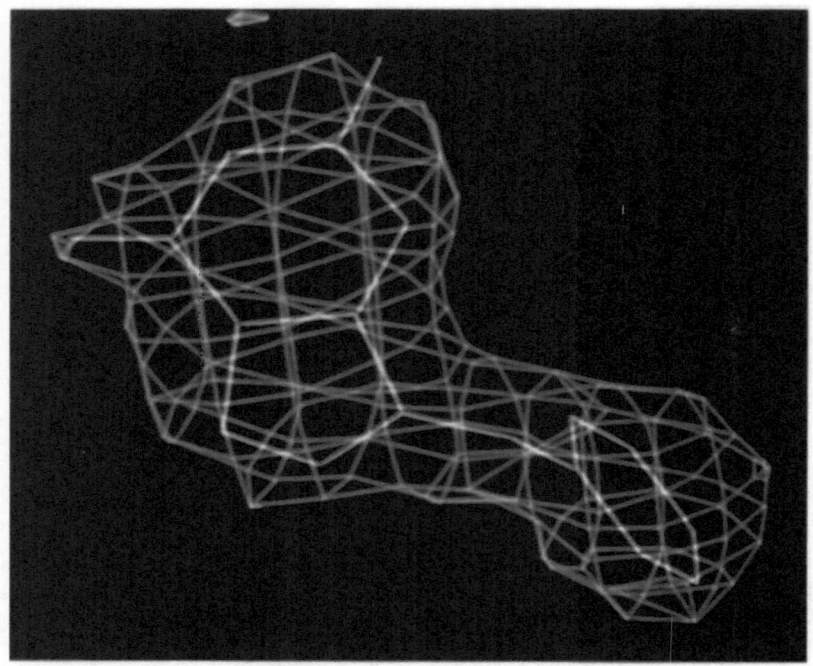

Figure 7 from "Design of Purine Nucleoside Phosphorylase Inhibitors Using X-Ray Crystallography," by Steven E. Ealick, Y.S. Babu, S.V.L. Narayana, William J. Cook, and Charles E. Bugg, p. 47.

Figure 10 from "Design of Purine Nucleoside Phosphorylase Inhibitors Using X-Ray Crystallography," by Steven E. Ealick, Y.S. Babu, S.V.L. Narayana, William J. Cook, and Charles E. Bugg, p. 50.

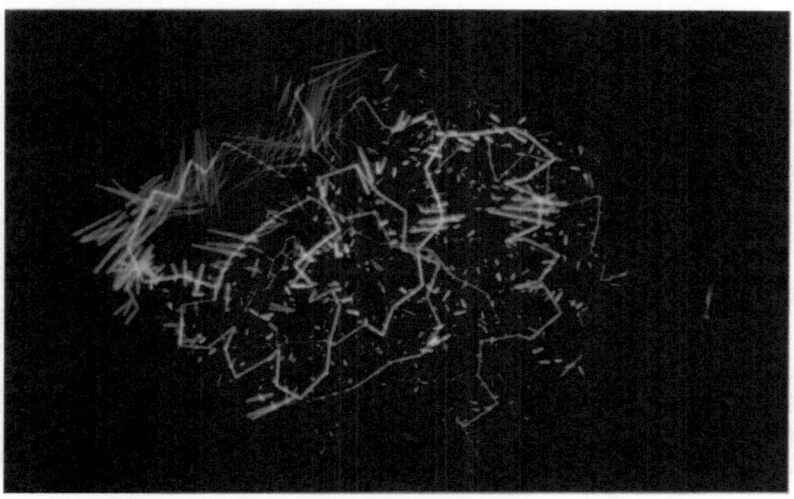

Figure 14 from "Design of Purine Nucleoside Phosphorylase Inhibitors Using X-Ray Crystallography," by Steven E. Ealick, Y.S. Babu, S.V.L. Narayana, William J. Cook, and Charles E. Bugg, p. 53.

Figure 3 from "Molecular Recognition of DNA Minor Groove Binding Drugs," by Andrew H.-J. Wang and Mai-kun Teng, p. 128.

a

b

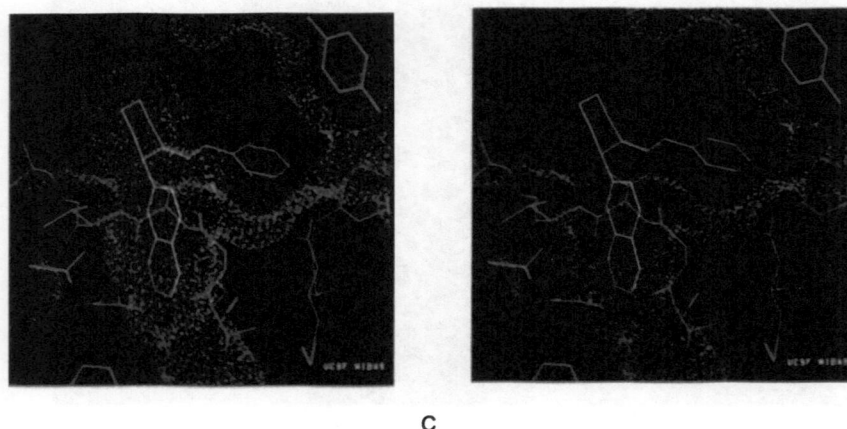

c

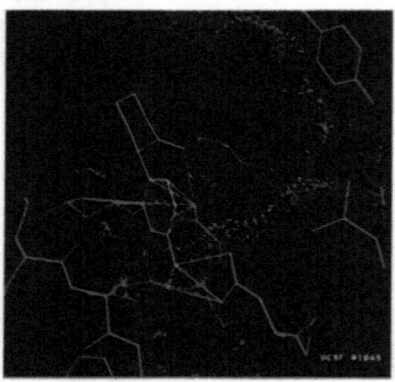

Figure 1 from "Inhibitor Design from Known Structures," by Renee L. DesJarlais, Brian Shoichet, George Seibel, and Irwin D. Kuntz, Jr., p. 204.

Structural and Computational Studies of Anticonvulsants: A Search for Correlation Between Molecular Systematics and Activity

PENELOPE W. CODDING, N.E. DUKE,
L.J. AHA, L.Y. PALMER, D.K. MCCLURG,
and M.B. SZKARADZINSKA

INTRODUCTION

Epilepsy is the most common form of neurological disorder except for stroke (Porter, et al., 1987). The disease is a collection of disorders, and a patient may suffer from more than one form of epilepsy. In addition, not all anticonvulsant drugs are effective against all types of seizure. There are two main classes of epileptic seizure, partial and generalized. These two classes of seizure are initiated differently: partial seizures begin at a localized focal point in the brain and spread to surrounding neurons while generalized seizures show no localized initiation point. Control of these two seizure types requires different activity: partial seizures can be controlled by preventing the sudden electrical discharge at the focal point and generalized seizures can be prevented by a general lowering of neuronal reaction.

This complex pharmacological picture is daunting for the application of drug design principles. A further complication arises when the target tissue for anticonvulsant drugs is sought, for, as yet, the sites of action for anticonvulsant drugs are unknown. Catterall (1987) has suggested that anticonvulsants can be separated into three classes on the basis of their activity both in *in vivo* tests and against specific seizure types and that these classes have different sites of action. This classification proposes that class 1 drugs, which only protect against maximal electroshock (MES) induced seizures, elicit their action by binding to a site on Na^+ channels, while class 2 drugs, which are effective against both MES and chemically induced seizures (subcutaneous Metrazol, scMET) and against a wide variety of seizures, act by facilitating the inhibitory neurotransmitter system mediated by γ-aminobutyric acid (GABA). The third type of anticonvulsant is no effective in both MES and scMET tests yet only shows protection against absence seizures; the mechanism for class 3 action is unknown. Catterall's findings suggest that *in vivo* tests can be used to identify anticonvulsants which may have similar modes of action, and thereby provides a beginning point for an analysis of molecular properties to seek structural principles for the development of new anticonvulsant drugs.

The studies presented herein are focused on MES protective drugs. The principle that guides this work is that analysis of structural, conformational and electrostatic features of drugs effective in the same *in vivo* test can provide insight into the required features of active drugs. Since no common target site has been proven, a general catalogue of the properties of anticonvulsants must be sought by building on the structural information currently available. The crystal structures of two classic MES drugs, diphenylhydantoin (Camerman, et al. 1971) and carbamazepine (Terrence, et al. 1983 and Himes, et al. 1981) have been reported. In addition, Codding, et al. (1984) surveyed the structures of

151

diphenylhydantoin   carbamazepine

five MES effective drugs and found that the conserved features in these drugs are: a planar amide group and a hydrophobic phenyl ring which are approximately perpendicular to each other. However, the crystal structure of diphenyloxazolidinedione (Codding, 1984) shows that these features are a necessary but not sufficient requirement for activity. The oxazolidinedione structure is similar but the compound is not effective against electroshock, suggesting that the electrostatic differences between a hydantoin ring and a oxazolidinedione ring are important.

Structural and computational studies on three new types of MES anticonvulsants are presented to test the pattern identified earlier and to suggest that a common molecular conformation and substitution pattern exists.

α-BENZYL GLUTARIMIDES

A series of halogen-substituted α-benzyl glutarimides were prepared by J. F. Wolfe and co-workers (Goehring, et al. 1983). The α-benzyl glutarimides are effective in protecting against MES induced seizures; the activity profile of the compounds is determined by the position and number of the halogen substituents on the benzyl ring, as summarized in Table 1. Either *ortho* or *para* halog+en substituents confer activity on the compound; *meta* substitution or di-substitution eliminates activity. The inactive N-methyl compound indicates that the N-H group is required for activity; a N-H proton donor is postulated to be a required functionality for MES anticonvulsants (Codding, et al. 1984).

This dependence on halogen substitution could be due to a non-bonded interaction between the phenyl ring π system and the dipole of the amide, as has been observed in the nmr spectra of benzyl-3-arylhydantoins (Fujiwara, et al. 1979). This interaction would be sensitive to the electronic changes induced in the rings when substituted in the *ortho* and *para* positions. Alternatively, the substituents may function by favoring a particular conformation of the phenyl ring relative to the glutarimide ring, thus facilitating either the non-bonded intramolecular interaction or a drug - receptor interaction. The structure-activity relationships do not distinguish between steric or electronic effects.

The crystal structures of the five compounds in Table 1 were determined to ascertain the effect of the substituent on conformation and electronic character. As shown in Figure 1, the five crystal structures show similar conformations for the phenyl ring relative to the glutarimide ring. In all cases, the aromatic ring adopts a conformation that is out of the plane of the glutarimide ring by more than $50°$. The o-Br compound has a nearly perpendicular arrangement of the two ring systems. Thus, in the crystalline state, no dissimilarity in conformation is found between the active and inactive compounds. The crystallographically determined bond distances show no systematic differences between the two classes of compound. To probe the electron-distribution in the molecules, semi-empirical molecular orbital calculations were done using the CNDO method (Pople, et al., 1966) and using the crystallographically determined geometry. These calculations indicate that a delocalization of electrons in the amide is characteristic of the active glutarimides. The active compounds have a stronger C_1-C_2 bond, a weaker

Table 1. Structures and activity profiles of the α-benzyl glutarimides.

Inactive:

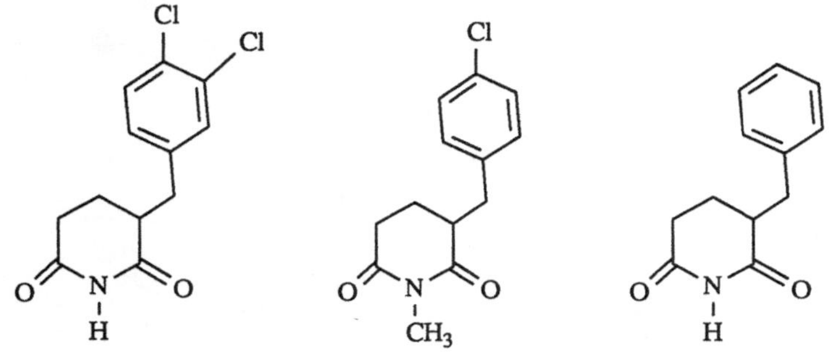

Active:

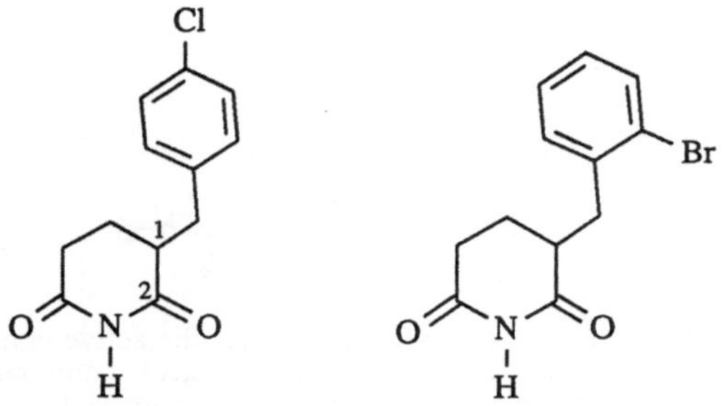

carbonyl bond (C_2=O), and a stronger N-C_2 bond; thereby indicating a stronger contribution from the ionic form of the glutarimide: $H\cdot\cdot N^+$=C_2-O^-. This trend enhances the proton donor role of the active glutarimide molecules, suggesting that in addition to any steric effects, the halogen substituents also facilitate the formation of a hydrogen bond at the recognition site for the anticonvulsant drug.

PHENYLPIPERDINOPYRIDAZINES

Pyridazine compounds have been useful as benzodiazepine receptor ligands, antidepressants and inhibitors of norepinephrine uptake. Based on the effectiveness of the pyridazine ring as a central nervous system agent, a series of pyridazine anticonvulsants were developed (Hallot, et al. 1986). The 6-aryl 3-(hydroxy polymethyleneamino) pyridazines have appreciable activity against MES induced seizures and show a potency comparable to class 1 anticonvulsants under the scheme due to Catterall outlined above. This series requires a 6-phenyl substituent for appreciable activity and a 4-hydroxypiperidine side chain in the 3-position of the pyridazine ring. Activity is also enhanced by the presence of an *ortho* chloro-substituent on the phenyl ring. This similarity to the structure-activity profile of the benzylglutarimides prompted a study of the structures of these ligands to determine the effect of the halogen atom. Table 2 summarizes the compounds studied.

The crystal structures of the five compounds are shown in Figures 2 and 3. Two notable features are evident: the *o*-Cl substituted phenyl ring is twisted out of the plane of the pyridazine ring more than the unsubstituted ring, and the conformation of the piperidine ring changes with the oxidation at the 4-position. The average angle between the planes

154

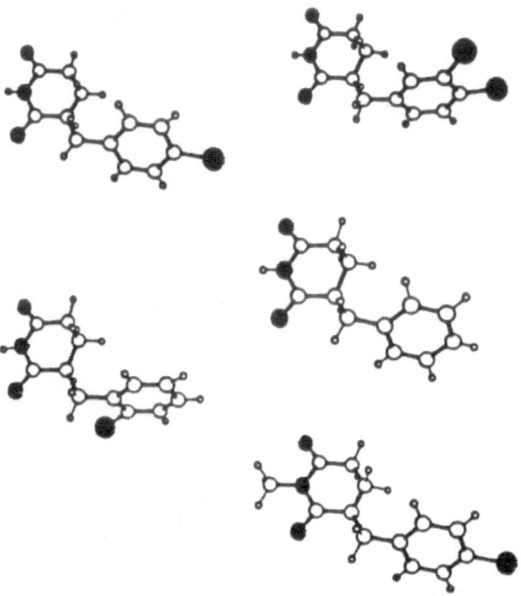

Figure 1. The crystal structures of the five benzylglutarimides: the active compounds, the 4-Cl and 2-Br derivatives, are on the left and the inactive compounds, the 3,4-dichloro, unsubstituted, and N-methyl 4-chloro derivatives are on the right.

of the phenyl and pyridazine rings for the four Cl-phenyl compounds (II-V) is 49.76° and the same angle for the unsubstituted phenyl ring (I) is 32.97°. This difference suggests that the increased activity of the o-Cl compounds may be related to the conformation of the phenyl ring. An o-Cl substituent may favor a more perpendicular arrangement of rings, like that found for other MES anticonvulsants (Codding, et al., 1984) and thus may contribute favorably to the entropy of binding by stabilizing the preferred conformation.

Table 2. The structures and pharmacological activity of the phenylpiperidinopyridazines studied.

Cpd.	R_1	R_2	R_3	R_4	ED_{50}(mg/kg)[†]
I	H	H	H	-OH	86
II	H	Cl	H	-OH	16
III	H	Cl	-OH	H	inactive
IV	H	Cl	H	=O	40
V	Cl	Cl	H	=O	5

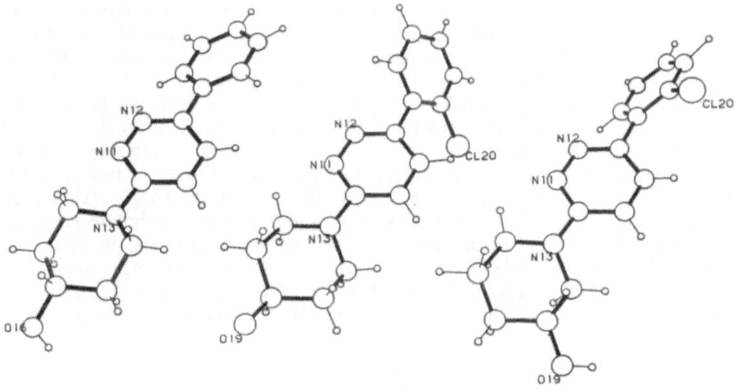

Figure 2. The crystal structures of the hydroxypiperidinopyridazines. The central 4-OH compound is active and the two flanking compounds are inactive.

To probe the effect of the *o*-Cl substituent, MOPAC calculations of the relative energies of the three hydroxy compounds, I, II, and III, were computed as the relative angle between the phenyl and pyridazine rings was stepped in 10° intervals. The results of the calculations indicate that the substituent affects both the rotational barrier for the phenyl ring and its conformation. The minimum energy conformations of II and III have the phenyl ring perpendicular to the pyridazine ring; and, the barrier to rotation is highest for a coplanar conformation: 8.93 and 9.11 kcal/mole, respectively. In contrast, compound I, which has low activity, has less than 1.0 kcal/mole variation in the energy as the phenyl ring conformation is varied and thus shows no preference for the position of the ring. MNDO calculations of the atomic charges do not support the hypothesis that the *o*-Cl substituent plays an electrostatic role: charge distributions in the pyridazine and phenyl rings were similar for all three hydroxypiperidine compounds.

The hydroxypiperidine compounds adopt an extended conformation while the two oxopiperidine rings adopt a bent conformation. This difference, or the replacement of a

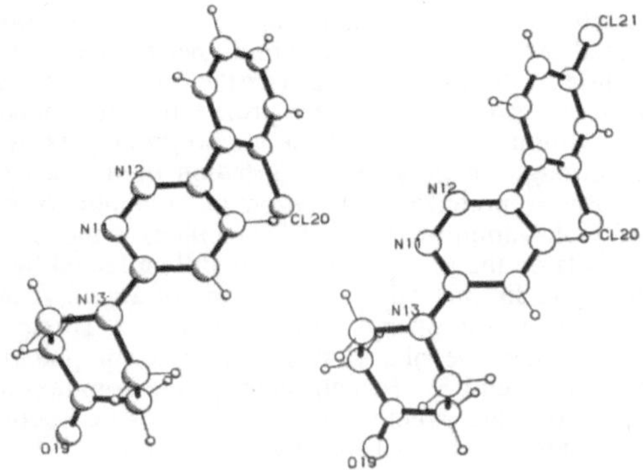

Figure 3. The crystal structures of the two oxopiperidinopyridazines. The piperidine adopts a distinctly different conformation from that found in the hydroxy compounds depicted in Figure 2.

hydrogen atom donor with a hydrogen atom acceptor, may be related to the increased toxicity of the two carbonyl-containing compounds. All five piperidine rings adopt a torsion angle about the C-N bond connecting the piperidine and pyridazine rings that allows interaction of the lone pair of electrons on the piperidine N atom with the π system of the pyridazine ring. This participation is evident from the bond distances in the pyridazine ring and from the molecular orbital calculations which show participation of the piperidine ring nitrogen atom in the π system of the pyridazine ring. The extended π system between the piperidine and pyridazine rings produces a polarization of the pyridazine N=N bond; the N atom closest to the piperidine ring carries the greater charge, -0.11 vs. -0.055. The average bond distances in the pyridazine ring, as shown below, support a delocalization of electrons between the two rings; in contrast, such a delocalization does not occur between the pyridazine and phenyl rings, due partly to the lack of coplanarity; MO calculations show that a parallel orientation of the phenyl ring in the 6-position disfavors the π interaction between the pyridazine ring and the piperidine N atom.

1.326(6)Å   1.348(4)Å   1.341(4)Å

Thus, in this system we find a common conformation for the active compounds that places two planar systems nearly orthogonal to one another. These compounds parallel the benzylglutarimide series and diphenylhydantoin by substituting the N=N of the pyridazine for the amide moiety common to other anticonvulsants. In these compounds, the third ring containing a 4-OH group is critical to activity, thereby suggesting that a hydrogen atom donor site is part of the anticonvulsant activity profile.

4-AMINO-N-PHENYLBENZAMIDES

A series of potent 4-aminobenzamide anticonvulsants have been reported by Clark and Robertson (Clark, et al., 1984, 1985, 1986, 1987a & 1987b and Robertson, et al., 1985 & 1987). These compounds consist of a linear arrangement of three planar systems and show MES activity; they are thus similar to the phenylpiperidinopyridazines. The central planar system in these compounds is an amide group; this feature is consistently found in MES anticonvulsants like diphenylhydantoin and carbamazepine.

Extensive structure-activity data have been reported on these compounds; these data show that for optimal MES activity, an amino-benzamide with an alkyl or aryl substituent on the amide N atom is required. For the amino substituent, MES activity is highest for a *para* amino group; the activity profile for compounds containing amino substituents at other positions suggests that the amino group acts by interacting with the receptor rather than through electrostatic perturbation of the rest of the molecule. A primary amino function is preferred; alkylation or acylation of the N atom reduces activity. MES activity of 4-amino-N-(phenyl)benzamides is increased by the addition of *ortho*-methyl substituents on the phenyl ring and is little affected by substituents in other positions. MES activity is diminished by the addition of a methylene group between the amide carbonyl and the amino-phenyl ring; however, a single methylene addition between the amide N-H and the phenyl ring produces compounds which have MES activity, for example compound 5. To help identify the common structural features of the benzamides and to correlate structure with activity for these compounds, the crystal structures of the benzamides shown in Table 3 were determined.

The crystal structures of compounds 1, 3, and 5, contain one molecule in the asymmetric unit; compound 4 has 2 molecules in the asymmetric unit and compound 2 has four. These nine observations of the molecular conformation of the anticonvulsant benzamides

Table 3. The structures and pharmacological activity of the benzamides studied.

Compound	R_2	R_3	R_6	MES ED_{50} (mg/kg)	scMET ED_{50} (mg/kg)
1	H	H	H	50.5	59.1
2	CH_3	H	H	~13.5	>300
3	H	CH_3	H	47.0	87.9
4	CH_3	H	CH_3	2.60	>300

Compound				MES ED_{50}	scMET ED_{50}
5				10.9	41.78

are shown in Figures 4 and 5. In all observations of the molecular conformation, the molecule exists as three planes which are not necessarily coplanar. In general, the amino-phenyl ring lies closest to coplanarity with the central amide plane: the average magnitude of the torsion angle between the amino-phenyl and the amide planes is $158 \pm 12°$; whereas the other phenyl ring adopts a wider range of positions and is closer to perpendicular to the central plane: the average magnitude of the torsion angle between the phenyl and the amide planes is $125 \pm 28°$.

Molecular mechanics (MMP2, Allinger, 1982) calculations were used to probe the barrier to rotation of the phenyl ring which has methyl substituents in compounds 2, 3, and 4. In compounds 1 and 3, which have an unsubstituted phenyl ring and a *meta*-substituted phenyl ring, respectively; the rotational barrier was the same height and the lowest energy structure had a torsion angle of *ca.* 30°. In contrast, the *ortho* substituted compounds showed the highest barrier for a coplanar arrangement of the phenyl ring and the amide group and the low energy structure had a torsion angle between the phenyl ring and the amide group of greater than 45°. Indeed, the addition of a second *ortho*-methyl group in compound 4, shifted the minimum energy structure to a inter-plane torsion angle of *ca.* 60° and raised the energy of the coplanar structure. These results suggest that the *ortho*-methyl substituents ·dramatically change the preferred conformation of the phenyl ring - amide portion of the molecule; therefore, the higher activity of compounds 2 and 4 may be due to a restriction of the molecular conformation to one that facilitates recognition. Compound 5 is found in the crystal to have a single conformation which places the phenyl ring perpendicular to the central amide group; thus, the high activity of this compound may arise from the intervening chiral carbon atom which favors a perpendicular conformation that may be required for activity.

The amino-phenyl ring and the central amide group adopt a more coplanar conformation in the crystal; this conformation is also found in molecular mechanics calculations which determined that the barrier to rotation about the amide to amino-phenyl linkage was 5 -

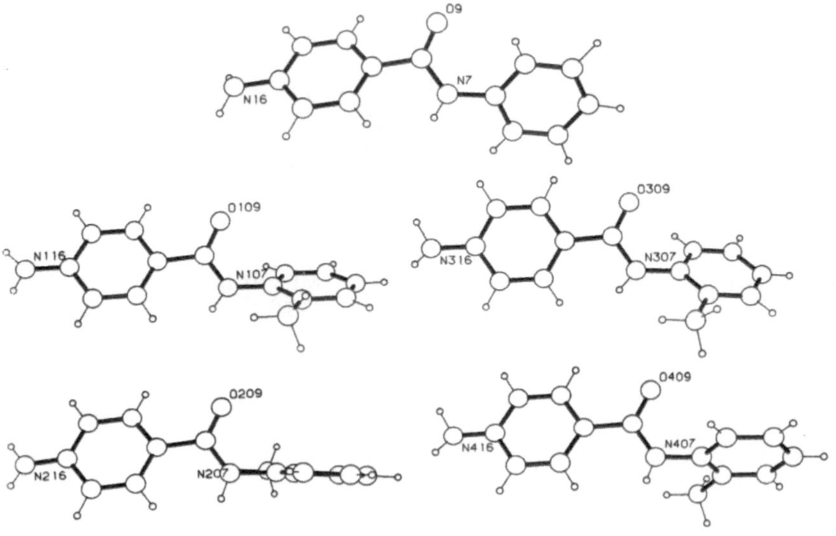

Figure 4. The molecular structures of amino N-(phenyl) benzamides **1** (top) and **2**. The four different conformations of compound **2** found in the crystal are shown at the bottom of the drawing.

12 kcal/mole higher than the barriers found in the *o*-methylphenyl to amide driver calculations. The minimum energy conformation for this portion of the molecule places the amide group and the amino-phenyl ring within 10° of coplanarity. This consistent conformation, found in both molecular mechanics calculations and the crystal structures,

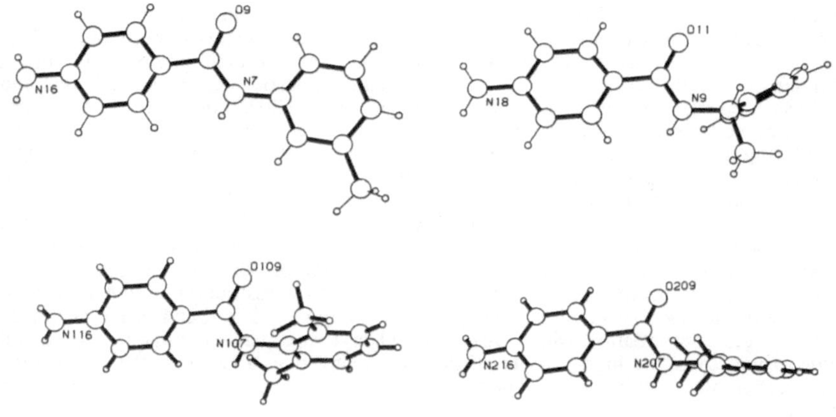

Figure 5. The molecular structures of amino N-(phenyl) benzamides **3** and **5** (the top line of the drawing) and the two conformations of compound **4** (at the bottom of the drawing).

and the decrease in activity resulting from addition of methylene group between the amide and the amino-phenyl suggests that the amino-phenyl ring and the amide group form an extended π system.

Thus, in this series of compounds we find that activity depends on the nearly perpendicular arrangement of two aromatic systems and the presence of a hydrogen bond donor at the opposite end of the molecule from the perpendicular *ortho*-substituted ring. In the crystals, the amino group participated in extensive hydrogen bonding thereby suggesting a mode of interaction with the recognition site for these drugs.

CONCLUSIONS

The consistent features found in these three classes of anticonvulsants are: the presence of a central group which has high electronegativity - the amide in glutarimides and benzamides and the N=N in the pyridazines - and the presence of an aromatic ring that is perpendicular to the central planar region. The pyridazines and the amino-benzamides have even greater consistency since they are of similar size and contain a third ring which carries a hydrogen atom donor group at the end of the molecule. The close parallel in the structural properties and conformational energy profiles of the pyridazines and the amino-benzamides suggests new modifications of both series of compounds. The common features of the structure-activity profiles of these two series of anticonvulsants also suggests that they share a common recognition site which requires a linear arrangement of a perpendicular phenyl ring (possibly with bulky *ortho* substituents); an electron donor atom (amide carbonyl or N=N double bond), and a hydrogen atom donor site.   This model suggests further modification of the benzylglutarimides to add the third recognition feature.

We have presented three examples of systematic crystallographic investigation of the structural features of drugs with similar activity profiles. These studies illuminate the consistent molecular features that are characteristic of active anticonvulsant drugs and the sum of the three investigations suggests new avenues for the design of MES active anticonvulsants.

ACKNOWLEDGMENTS

We thank the Alberta Heritage Foundation for Medical Research (studentship for N.E.D., L.Y.P., D.K.M. and scholarship for P.W.C.) and the Medical Research Council of Canada (grant MA8087 to P.W.C.) for financial support. We also thank our collaborators J.F. Wolfe, C.R. Clark and D.W. Robertson for generously supplying samples for our studies.

REFERENCES

Allinger, N.L. 1982. MMP2: Molecular Mechanics, Quantum Chemistry program Exchange, Chemistry Department, Indiana University, Bloomington, Indiana 47405, USA.

Camerman, A. and N. Camerman. 1971. The stereochemical basis of anticonvulsant drug action.   The crystal and molecular structure of diphenylhydantoin, a noncentrosymmetric structure solved by centric symbolic addition. Acta Crystallogr., Sect. B, 27:2205-2211.

Catterall, W.A. 1987. Common modes of drug action on Na⁺ channels: local anesthetics, antiarrhythmics and anticonvulsants. Trends in Pharmacol. Sci., 8:57-65.

160

Clark, C.R., M.J.M. Wells, R.T. Sansom, G.N. Norris, R.C. Dockens, W.R. Ravis. 1984. Anticonvulsant activity of some 4-aminobenzamides. J. Med. Chem., 27:779-782.

Clark, C.R., R.T. Sansom, C.M. Lin, G.M. Norris. 1985. Anticonvulsant activity of some 4-aminobenzanilides. J. Med. Chem. 28:1259-1262.

Clark, C.R., C.M. Lin, and R.T.J. Sansom. 1986. Anticonvulsant activity of 2 and 3 aminobenzanides. J. Med. Chem. 29:1534-1537.

Clark, C.R., and T.W. Davenport. 1987a. Synthesis and anticonvulsant activity of analogues of 4-amino-N-(1-phenylethyl) benzamide. J. Med. Chem. 30:1214-1218.

Clark, C.R., T.W. Davenport. 1987b. Anticonvulsant activity of some 4-aminophenyl acetamides. J. Pharmceut. Sci. 76:18-20.

Codding, P.W. 1984. The structure of 5,5-diphenyl-1,3-oxazolidine-2,4-dione, $C_{15}H_{11}NO_3$. Acta Crystallogr. Sect. C, 40:2071-2074.

Codding, P.W., T.A. Lee and J.F. Richardson. 1984. Cyheptamide and 3-hydroxy-3-phenacyloxindole: structural aimilarity to diphenylhydantoin as the basis for anticonvulsant activity. J. Med. Chem., 27:694-654.

Fujiwara, H., A.K. Bose, M.S. Manhas, and J. M. van der Veen. 1979. Non-bonded aromatic-amide attraction in 5-benzyl-3-arylhydantoins. J.C.S. Perkin II, 653-658.

Goehring, R.R., J. Subrahmanyam, and J.F. Wolfe. 1983. Abstract MEDI 58, 186th ACS Meeting, Washington, D.C.

Himes, V.L., A.D. Mighell and W.H. De Camp. 1981. Structure of carbamazepine: 5H-dibenz[b,f]azepine-5-carboxamide. Acta Crystallogr., Sect B, 37:2242-2245.

Pople, J.A. and G.A. Segal. 1966. CNDO. J. Chem. Phys. 44:3289-3295.

Porter, R.J. and W.H. Pitlick. 1987. in "Basic and Clinical Pharmacology", 3rd edn., Katzung, B.G., ed., Norwalk: Appleton and Lange, pp. 262-278.

Robertson, D.W., E.E. Beedle, J.D. Leander, and R.C. Rathburn. 1985. Comparison of the anticonvulsant activities of the R[LY(R)] and S[LY(S)] enantiomers of 4-amino-n-(α-methylbenzyl)-benzamide. Pharmacologist 27:231.

Robertson, D.W., J.D. Leander, R. Lawson, E.E. Beedle, C.R. Clark, B.D. Potts, C.J. Parli. 1987. Discovery and anticonvulsant activity of a potent metabolic inhibitor 4-amino-N-(2,6 dimethylphenyl) 3,5-dimethylbenzamide. J. Med. Chem. 30:1742-1746.

Terrence, C.F., M. Sax, G.H. Fromm, C.-H. Chang, C.S. Yoo. 1983. Effect of baclofen enantiomorphs on the spinal trigeminal nucleus and steric similarities of carbamazepine. Pharmacol., 27:85-94.

Crystallography and Molecular Mechanics in Designing Drugs with Unknown Receptor Structure

DAVID J. DUCHAMP

Introduction

This paper presents an approach to bringing three-dimensional information from crystallography and molecular mechanics into the drug design process in a timely and effective manner. The case treated is the one where the three-dimensional structure of the drug receptor is unknown, which unfortunately is still the case for most drug design problems. The approach presented is not unique, but is one that has worked and one that embodies many years of experience. Other papers in this volume discuss the more ideal case where the receptor structure is known. Some of the comments presented here apply to that case also.

The goal in employing molecular mechanics and crystallographic information in a drug design project is to construct a three-dimensional representation of the molecular shape and location of features required for a molecule having a certain drug action. This three-dimensional information and other information, such as, biological activity data, is used to find the requirements for a drug molecule binding at a specific receptor of unknown three-dimensional structure. This may appear to be an impossible task, but experience is that even if the receptor structure is unknown, quite a bit of progress toward this goal can be made.

Such work is used to gain insight into requirements for drug action, and thereby to speed up lead development and to help in design of molecules with the desired action and without undesired features (side effects). The key word is "help." This type of work usually works best as a collaboration between structural physical chemists, who understand and can accomplish the experiments and calculations involved, and medicinal chemists, who understand and can accomplish the synthetic side of the problem. The most unproductive arrangement is when the structural chemist, working alone, designs "ideal" molecules and attempts to dictate which molecules are to be synthesized by others. Most medicinal chemists, working alone, will lack a deep insight into the strengths and limitations of the calculations and experiments, and will be less capable of deciding which experimental or calculational approach will be best suited for the next step in the process.

Design of a drug usually goes through several steps, including: development and set up of biochemical and/or biological tests for evaluation of activity, testing many molecules until a "lead" compound is discovered, preparing many analogs to a "lead" molecule to improve activity, and detailed evaluation of the most promising leads for side effects, toxicity, bioavailability, and metabolism. Where the receptor is unknown, three-dimensional structural methods may be employed at any phase in this process, however, they are usually more effective after biological data is available on a series of molecules which exhibit the same desired pharmacological endpoint. Also these methods work better if data is available for molecules with more than one basic structure. This type of work tends to be most effective when a new lead has been discovered, and an analog program is in the planning stage or has just begun.

Discussion

Below are listed the basic steps necessary for this design process. Some researchers will add an additional step, or combine some of the ones mentioned here. This list of steps is not unique, but does outline the process of using three-dimensional information in the drug design process where the receptor structure is unknown. Each step in this list will be discussed in more detail during the remainder of this paper.

1. Assemble and review known data.
2. Evaluate experimental results.
3. Test the applicability of the molecular mechanics force fields to the molecules to be studied.
4. Use molecular mechanics to explore conformational flexibility.
5. Use molecular graphics to gain insight into possible common features of molecules.
6. Use co-minimization of molecules to test ideas of common three-dimensional features.
7. Propose "active" conformations.
8. Develop a "model" for the three-dimensional requirements for binding.
9. Collaborate with medicinal chemist in suggesting, making, and testing synthetically feasible analogs to explore certain features of the "model."
10. Modify the "model" based on results of tests.

The order of these steps is important when attempting to apply an organized scientific approach to studying the three-dimensional requirements of a receptor of unknown structure. Taking them out of order can lead to failure, or at best, useless work. For example, jumping directly to step 7, guessing an active conformation, can lead to much useless work where researchers try to justify their initial assumption, instead of systematically approaching the problem. An example of this type of side track is when a crystallographer assumes from the outset that the conformation found for the molecule in the crystal must be the active conformation, and then tries to "prove" it. Although the steps should be approached in order, there is usually quite a bit of reiteration back to previous steps during the process.

Assemble and review known data (1)

Two types of data always need to be assembled and reviewed--three-dimensional structural data and pharmacological data. In addition for certain drugs, bulk physical chemical properties, such as, solubility, pK, and hydrophobicity, can be very important. The stability of the drug molecule in various media, and its partition coefficient are other properties affecting drug action. Previous structure/activity relationship (SAR) work, if available, should also be reviewed, especially any formal QSAR studies which might have been published.

Pharmacological data classifies into two basic types: *in vivo* data from experiments in live animal model systems, and *in vitro* data from experiments performed in a model environment outside a living body. *In vitro* data usually give a better correlation with structural or physical chemical parameters than *in vivo* results, however, *in vivo* data are usually more pertinent to the final desired drug action. *In vivo* drug action is a complicated process. It is influenced by a number of factors including: drug absorption (in the case of non-IV administration), drug distribution, drug metabolism, receptor binding, secondary mechanisms for control of levels of endogenous substances, and mode and kinetics of drug excretion. Most structure activity relationships attempt to relate structural and physical chemical properties to the receptor binding phase of the drug action process. Receptor binding may or may not be the factor which is limiting the *in vivo* activity of a particular drug relative to its analogs. Whatever is known about the entire *in vivo* pharmacology of a drug should be reviewed in order to make effective correlations. In reviewing pharmacological results, it is important to look at inactive compounds as well as active compounds--both are needed for a good model.

Structural data come from experimental results and from results of calculations. Experimental three-dimensional structural results come primarily from crystal structure determinations on small molecules. In selecting which crystallographic results to study in detail, preference should be given to low temperature crystal structure work. Experimental conformational data from NMR should also be surveyed. The NMR experiments producing useful conformational results are almost always performed in solution; the solvent system used should be considered in evaluating the relevance of solution NMR results. Calculational results come primarily from molecular mechanics and quantum mechanics. In assembling and reviewing structural results from previous calculations, understanding the hypotheses and assumptions involved in the calculations is very important, as is a detailed review of the thoroughness of the work.

Evaluate Experimental Results (2)

Pharmacological data is extremely important, but must be evaluated very carefully, and with a high level of scientific skepticism. Pharmacological systems are very complicated, and pharmacologists almost always do not know exactly what they are measuring. Over the years the interpretation of previous pharmacological results tends to change as more is learned about these complex systems. Assays which are thought to be measuring certain specific end points, are sometimes found on further study to be addressing something different. Also *in vivo* results are sometimes not reproducible within the stated error limits, due usually to low numbers of animals used or lack of control of some unknown factor affecting the experiment. This is not a criticism of pharmacologists--some of the best scientists I know are pharmacologists--but is a reflection on the fact that pharmacology is a very difficult area of study.

Likewise structural results must be evaluated very carefully. Crystal structures provide a low energy conformation for the molecule, but not necessarily the "active" conformation. For drug-type molecules, unless extensive hydrogen bonding is present, a molecule will almost always crystallize in a conformation which is within 1 to 2 kcal/mole of the global minimum energy conformation for the free molecule. Thermal vibrations can have important effects on detailed crystal structure results, and must be taken into consideration when surveying and comparing molecular parameters. Anomalous or unexpected structural features--shorter or longer distances, etc.--can supply useful insight into structural characteristics of molecules. Part of the evaluation should be to decide if more crystal structure determinations are necessary. The final result of the evaluation process is to select series of molecules for further study.

Test the Applicability of Molecular Mechanics Force Fields (3)

Since not all molecular mechanics force fields are equally suitable for certain classes of molecules, the applicability of a given molecular mechanics force field to the series of molecules selected for further study needs to be tested. This can be done most effectively by comparing the results of molecular mechanics calculations using that force field to experimental results. We have found from much experience, that to effectively compare crystal structure experimental results with results of molecular mechanics calculations requires some form of "in-crystal" minimization. At the level of accuracy necessary for such comparisons, the influence of neighboring molecules in the crystal on the conformation of a given molecule is usually very significant. If an "in-crystal" calculation is not done, many force field defects may be explained away by discounting differences between observed and calculated values of molecular parameters to "crystal packing effects."

The CONFOS molecular mechanics program, developed by the author, uses a simulated crystal environment when minimizing crystal structures (Duchamp,et.al., 1988). In doing this the program: automatically generates all symmetry-related atoms within a user-specified distance, preserves the space group symmetry during minimization, holds unit cell parameters fixed during minimization, and also provides for the restriction of coordinates of atoms in special symmetry positions. During minimization, the positions

Table 1. Source of the 460 experimental structures used to evaluate the force field of the CONFOS Program (April, 1989).

No. of Structures	Experimental Technique
217	X-ray Diffraction
42	Neutron Diffraction
122	Microwave Spectroscopy
71	Electron Diffraction
4	Infrared Spectroscopy
3	Raman Spectroscopy
1	Ultraviolet Spectroscopy

of all atoms in the crystallographic asymmetric unit are varied, with their symmetry-related counterparts following along. Such a scheme does not produce good lattice energies, but does provide an effective representation of external forces from neighboring molecules.

Tables 1 and 2 give data which show the level of agreement which one might expect when comparing experimental results with molecular mechanics calculations. Table 1 shows the source of the 460 structures used to evaluate the force field of the CONFOS program, and table 2 gives the overall agreement with experiment currently achieved by this molecular mechanics program.

The final part of this step is to revise potential parameters as necessary to achieve the desired accuracy of agreement. When doing this, it is important to note that molecular parameters of functional groups in very small molecules often are not good predictors for the corresponding group in larger molecules of drug size. Formamide, for example, is not a good model for the amide bond in a peptide. The balance of the force field must be maintained when potential parameters are changed to compensate for poor agreement.

Table 2. Overall agreement between observed and calculated molecular parameters (April 1989). For each structural parameter, the number of types (potential parameter sets), the total number of observations, and the average differences between experimental and calculated values are given. Two averages are given: in the "type" average, the differences for each type were averaged together, then an average was taken over the type averages; the "observations" average is a straight average over all observations.

	Number		Average Difference	
	Types	Observations	Types	Observations
Bond Angles	577	9171	1.2°	1.3°
Bond Distances	351	6548	0.010 Å	0.012 Å
Torsion Angles	336	10764	3.8°	3.1°
Out-of-Plane	22	897	0.013 Å	0.012 Å
Hydrogen Bonds	89	1102	0.097 Å	0.080 Å
Non-Bonded	54	6755	0.090 Å	0.060 Å

Use Molecular Mechanics to Explore Conformational Flexibility of Molecules (4)

One of the most valuable studies that can be accomplished by molecular mechanics is to explore the structural limits for low energy conformations. Such a study is most meaningful if done systematically, first studying the molecule to ascertain possible areas of flexibility, then using molecular mechanics to test each possible flexible area. Using a "seat-of-the-pants" approach can lead to mistaken assumptions, usually assumptions that a molecule or part of a molecule is more rigid than it really is.

To explore conformational flexibility, adding "extra potentials" or "forcing potentials" to the usual strain energy expression is especially useful (Duchamp, 1977). This method allows for the facile application of one or more constraints on a molecule to allow study of conformations with other than the minimum energy. In this approach, the total energy, E_{total}, is minimized where

$$E_{total} = E_{strain} + \sum_{\substack{extra \\ potentials}} E_{extra}$$

and E_{extra} is an extra potential energy formulated to force a given conformational parameter toward a certain value. For example, a suitable extra potential for forcing a torsion angle, ω, toward a certain value, ω_o, would be

$$E_{extra} = c(\omega - \omega_o)^2$$

where c is a constant set to a large enough value to force the torsion angles to match within a certain difference.

Figure 1 illustrates a calculation in which a steroid, 19-norandrostenediol, has been tested for flexibility about the torsion angle shown in the drawing. The calculation was performed with the CONFOS program by using an option which introduces an extra potential, as shown above, and completely minimizes the entire molecule at regularly spaced values of ω_o to give the points shown on the graph. This torsion angle may be varied by about 20° without raising the conformational energy by 0.5 kcal/mole.

Use Molecular Graphics to Gain Insight into Possible Common Features of Molecules (5)

Once the conformational flexabilities of all the molecules are understood individually, the next step is to search for common features between molecules. Any previous SAR data can be helpful in providing starting ideas. Molecular graphics is the computer technique

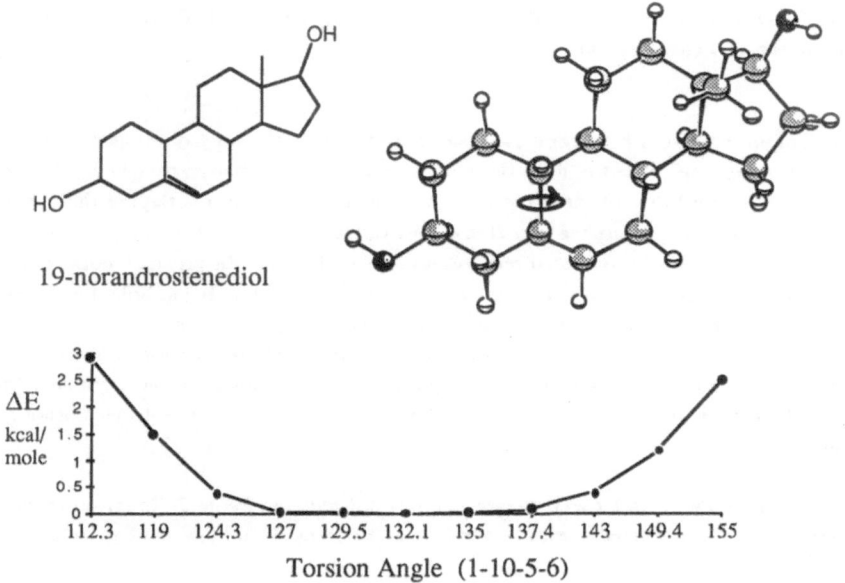

19-norandrostenediol

Torsion Angle (1-10-5-6)

Figure 1. The conformational flexibility of 19-norandrostenediol to rotation about the (1-10-5-6) torsion angle. The torsion angle is indicated in the ball and stick drawing of the molecule. The values plotted are torsion angle in degrees, horizontally, and relative conformational energy in kcal/mole, vertically.

most used here; in particular, molecule overlays are very useful. Surface or space filling displays can also be very helpful. Molecular models, both the space filling CPK type and the Dreiding type, are a useful supplement to computer graphics.

In comparison studies such as these, both active and inactive molecules must be studied. Only by study of inactive or weakly active molecules can insight be gained about undesirable features in candidate drug molecules. In cases where a drug molecule has one or more asymmetric sites, it is crucial in some types of activity to ensure that the correct enantiomer of the molecule is being studied. Enantiomeric specificity is especially important in drugs of the central nervous system and in drugs which are derivatives or mimics of optically active endogenous substances. For many drugs, one enantiomer possesses all the desired activity. In almost all drugs, there is some difference in activity between enantiomers. It is sometimes necessary to consider a racemate, either because there is no activity difference between enantiomers, or because there is no data on which enantiomer has the greater activity. Studying a racemate, requires that two molecules--both enantiomers--be studied separately. Most molecular graphics programs allow for simple switching between enantiomers.

Use Co-minimization of Molecules to Test Ideas of Common Three-dimensional Features (6)

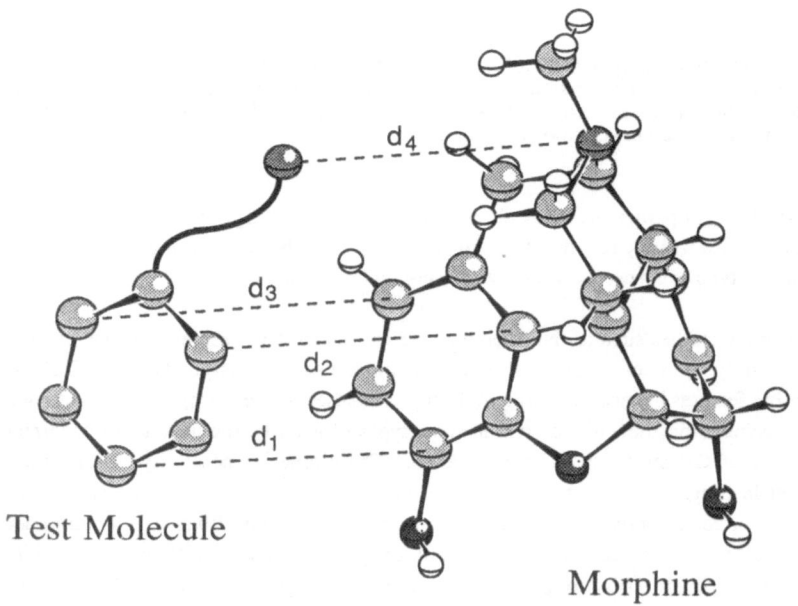

Test Molecule

Morphine

Figure 2. Illustration of the co-minimization of a test molecule to the morphine molecule at two sub-structure contacts at three points d_1, d_2, and d_3, and a fourth at a common point d_4.

Table 3. Results of co-minimization calculations of the named compound and morphine. Each calculation was a two-way (4-point) match with morphine. The underlined values are indicative of a poor fit. The distances correspond to those illustrated in figure 2.

Compound	Fit (Å) Aromatic	N	Energy Cost (kcal/mole) ΔE(Cmpd)	ΔE(Morp)
U-50488	0.13,0.03,<u>0.45</u>	<u>0.81</u>	<u>5.3</u>	0.3
U-47700	0.11,0.05,0.17	0.22	1.2	0.5
Cyclazocine	0.04,0.01,0.05	0.06	0.1	0.1

at the basic nitrogen. The extra potential formulation for this match would be

$$E_{extra} = c\left(d_1^2 + d_2^2 + d_3^2\right) + c'd_4^2$$

where different numerical values for c and c' would normally be used in order to control the tightness of spatial matching at the two sites in the minimized molecules. This formulation is the mathematical equivalent of zero length springs with different force constants. In this particular case the aromatic match should be tighter than the match at the basic nitrogen, because the basic nitrogen will be protonated at the site of action, giving an electrostatic interaction with a putative receptor, whereas the interaction of the aromatic moiety with a receptor will require tighter spatial constraints.

Besides obtaining molecules in "common" conformations, these calculations yield two types of numerical information which are useful in evaluating the feasibility of the proposed match. These are the actual distances achieved in the match and the increase in strain energies (relative to the corresponding fully minimized free molecules) necessary for the molecules to achieve the final conformations. Table 3 shows representative results from this type of calculation from some work we did on analgesics which bind at the μ opiate receptor. From the results shown in this table, one would conclude that cyclazocine matches morphine perfectly, U-47,700 matches quite well, but U-50,488 gives a poor match. The relative energy, 5.3 kcal/mole, needed by the U-50,488 molecule to achieve a matching conformation is too large, and the matching distances at one of the aromatic sites and at the basic nitrogen are too large also. Obviously these results have to be treated in a somewhat qualitative way, even though they are quantitative, because the force constants in the extra potential formulation can be set large to achieve an excellent spatial match at the expense of energy, and vice versa. In our experience, the force constants are properly set when the spatial distances and the relative energy deviate from feasible values simultaneously.

Propose "Active" Conformations (7)

By study of selected active and inactive molecules using molecular graphics and molecular mechanics, sufficient insight may be gained to allow proposal of an "active" conformation

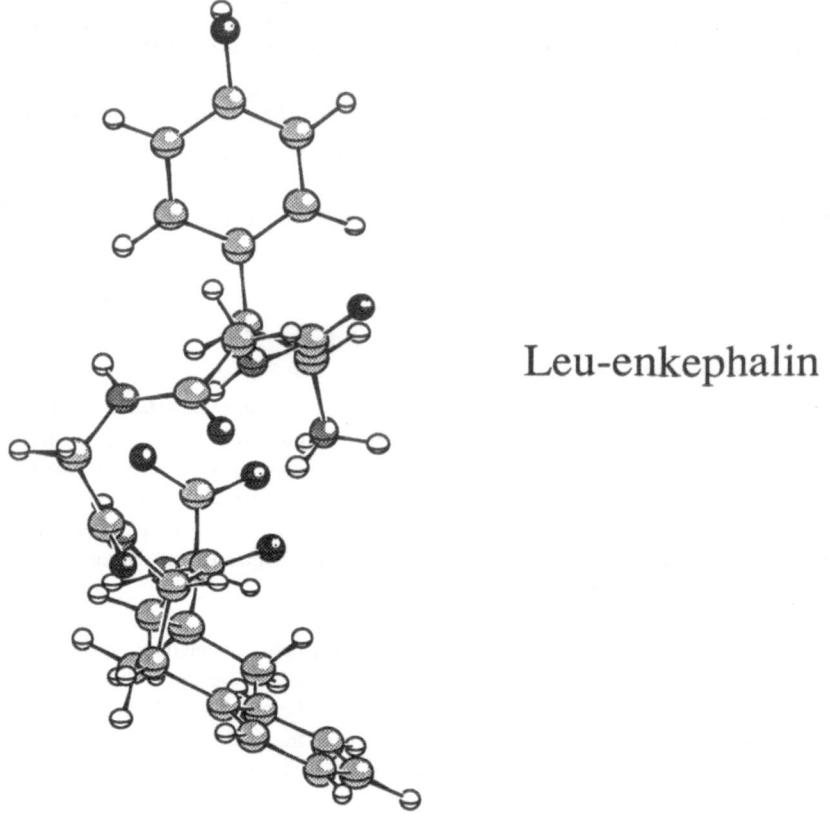

Leu-enkephalin

Figure 3. Suggested active conformation for the endogenous opiate peptide, leucine-enkephalin.

for each of the molecules being studied. These proposals may be tested and refined by reiterating with steps 5 and 6 above. It is important to ensure that all active molecules can assume the proposed conformations with reasonable accuracy at reasonable energy cost. Inactive molecules may or may not be capable of achieving a conformation matching the active molecules. Their inactivity could be due to a conformational problem or to a problem in a part of the molecule other than those needed to achieve an "active" conformation. Figure 3 illustrates a proposal I made for the active conformation of the endogenous opiate peptide, leucine enkephalin. This proposal, made within a few months of the announcement of the discovery of the enkephalin molecule (Hughes, et.al, 1975), was based on analgesic calculations we were making during that period.

Develop a "Model" for Three-dimensional Requirements for Binding (8)

Using the results derived from the above steps and all available pharmacological activity data, a model may be developed for the three-dimensional requirements for binding at the receptor. This model may embody various types of information, and its completeness and detailed specificity will vary depending on how many molecules of differing structural types can be brought to bear, and on the completeness and accuracy of the pharmacological data. The model may contain specific three-dimensional sites which must be occupied by specific types of moieties, e.g., a hydrogen-bond acceptor, or a basic nitrogen. Some sites may be required as "essential for activity", whereas some may be optional, making a "binding enhancing" contribution when present. The model may also contain qualitative three-dimensional qualifiers, such as, "the receptor will not accommodate any bulky substituents in this (specific three-dimensional) area," or "this side of the receptor requires hydrophobic contacts." The study of inactive or weakly active molecules which can achieve the "active" conformation frequently leads to this type of information. In developing a model, the absence of information bearing on certain areas of the putative receptor should also be noted, since this information can help guide future development.

The model should be evaluated against all existing relevant pharmacological data. The model should "explain" the qualitative ranking of molecules according to activity. Frequently, however, there will be some exceptions to the agreement with activity data for specific molecules. Such disagreements do not necessarily mean that the model is invalid, since the numerical value of the pharmacological result and the interpretation of what this result actually means are not immutable. Such disagreements can also suggest that some factor other than binding to the specified receptor may be influencing the activity of the specific molecule in question.

Figure 4 shows the spatial part of a three-dimensional model of the opiate μ receptor which was developed using these techniques. For comparison purposes a drawing of 3-methyl fentanyl in an "active" conformation corresponding to the model is also shown. The positions of the points in the model were derived by averaging the positions of the corresponding atoms in a large co-minimization calculation employing the most active opiate drug molecules. The model was then checked by minimizing a large number of opiate drug molecules against the points of the model (which were not varied). These calculations were performed by separately minimizing each of the molecules with extra potential distance links to the relevant fixed points of model. The quality of each match to the model was judged by evaluating the matching distances and the relative energy of the resulting matched molecule as mentioned previously for the co-minimization calculation.

As a model is developed and verified against known molecules, it may be necessary to refine it by reiterating through steps 5, 6, and 7 above. Once such a model is developed and verified, a proposed description of the requirements of the receptor is available in a form suitable for easy and quick match testing of molecules.

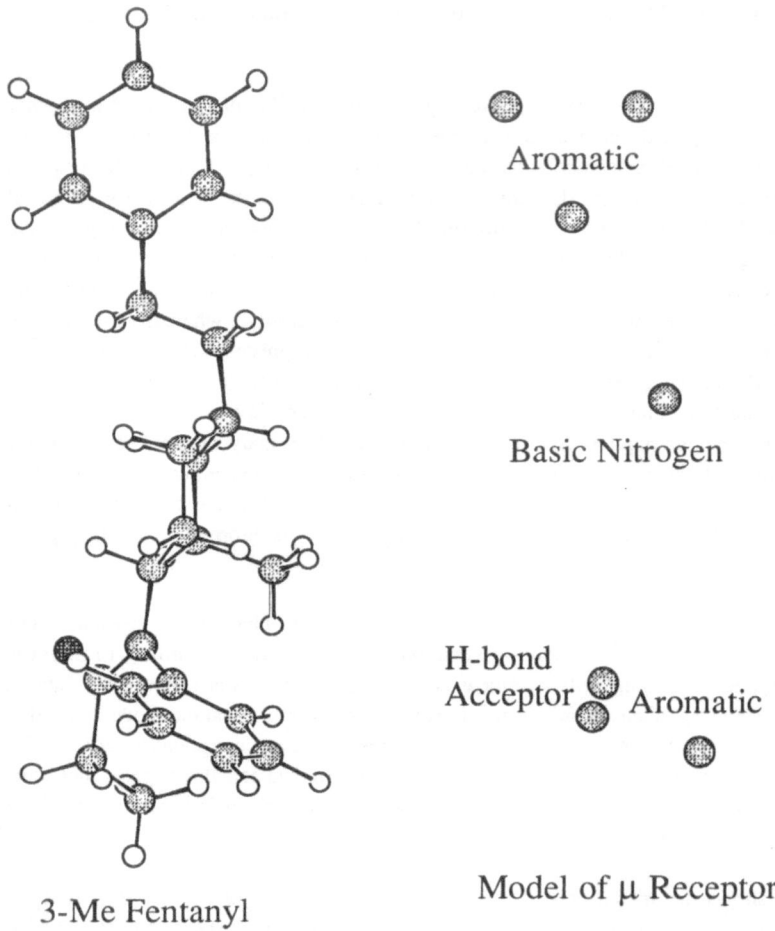

Figure 4. Seven-point three-dimensional model for the opiate μ receptor. 3-methyl fentanyl matched to the model is shown on the left. The upper aromatic and the basic nitrogen sites are essential for activity. The lower aromatic site and the hydrogen bonding site are optional.

Collaborate with Medicinal Chemist in Suggesting, Making, and Testing Synthetically Feasible Analogs to Explore Certain Areas of the Model (9)

Once a model has been developed, it can be used to help design new drug molecules. This can be done in several ways. A series of molecules being considered for synthesis may be rank ordered according to compliance with the model. Study of the model may suggest an area in three-dimensional space which could be explored synthetically. Comparison of an existing molecule with the model may suggest a site on the molecule which might be modified to enhance activity. This part of the study should always be done in collaboration with a synthetic chemist, who knows what is easily achievable synthetically, and who will be familiar with the patent literature.

When using a model of this type to aid in the design of drugs, one must be very careful not to over interpret the model. The model serves only to help in the design by stimulating three-dimensional ideas for the scientists involved. The strengths as well as weaknesses of the model must be kept in mind in formulating reasonable questions. Some correlations between structure and activity will depend upon molecular features not representable by shape or conformational energy.

If possible, when selecting molecules for synthesis, select molecules which will test the model and/or have a potential for extending the model. Such a test might be suggested, for example, when the model is incapable of differentiating between two or more molecules in predicted activity. Molecular graphics is also very useful in helping with visualization of the model and how it relates to molecules in three-dimensions.

Modify the "Model" Based on Results of Tests (10)

One is sometimes asked to design an experiment which will be definitive in either verifying or rejecting a model. The flaw in this request is that a model which is carefully built up as mentioned above, is never "rejected." When a deficiency is found in such a model, the model is modified and/or extended to correct the deficiency. Models are empirical by nature; they are never complete; and they are usually incapable of quantitative predictions, except over very restricted series of molecules. One expects to need to refine a model to improve its abilities to correlate known results and to predict the results of future experiments. Indeed, model refinement is a method of systematizing and organizing new structure/activity information. For a model to be useful, however, it must be capable of reasonably reliable qualitative predictions. This ability is improved by modification and extension of the model as new results are available. New knowledge of the pharmacology of the drug receptor may also supply insight allowing model improvement.

Conclusions

A procedure has been described for developing insight about the requirements for binding to a receptor of unknown three-dimensional structure. This insight derives from use of

174

three-dimensional structural data from crystallographic experiments and from molecular mechanics calculations to systematically build up a three-dimensional model for the requirements that need to be satisfied by a molecule binding at this receptor. Studies of this type can help in the design of drugs by providing an additional tool to help in the ranking of molecules being considered for synthesis, and by suggesting areas of a molecule that should be considered for modification. Effective use of these techniques usually requires a collaboration between a structural chemist and a medicinal chemist. These studies are most profitable if employed during lead development, beginning early in the process. The systems most amenable to this approach are those in which receptor binding is the factor most limiting drug activity.

Acknowledgements

The author gratefully acknowledges the help of Loraine M. Pschigoda in much of the work which formed the basis for this paper. Likewise the author is appreciative of the contributions of many colleagues at The Upjohn Company, who have freely given advice and consultation and participated in collaborations which led to many of the ideas mentioned above.

References

Duchamp, D. J. 1977. "Newer Computing Techniques for Molecular Structure Studies by X-ray Crystallography," in *Algorithms for Chemical Computations* (ed. R. E. Christoffersen) pp 98-121. American Chemical Society, Washington.

Duchamp, D. J., Pschigoda, L. M., and Chidester, C. G., 1988. "Molecular Mechanics, Crystallography, and Drug Research," in *Molecular Structure, Chemical Reactivity and Biological Activity* (ed. J. J. Stezowski) pp 34-39. Oxford University Press, New York.

Hughes, J., Smith, T. W., Kosterlitz, H. W., Fothergill, L. A., Morgan, B. A., and Morris, H. R., 1975. "Identification of Two Related Pentapeptides from the Brain with Potent Opiate Agonist Activity." Nature 258(5536): 577-579.

Molecular Modeling with Substructure Libraries Derived from Known Protein Structures

Barry C. Finzel, S. Kimatian, D.H. Ohlendorf,
J.J. Wendoloski, M. Levitt, and F. Ray Salemme

ABSTRACT

Many studies have illustrated that proteins are hierarchical assemblies of localized substructures. Here we describe the organization and use of an interactive computer program that allows the graphical construction of protein models using fragments extracted from known x-ray crystal structures. An initial α-Carbon conformational template is used to screen and retrieve matching polypeptide fragments from a library of known protein structures. Fragments are evaluated either graphically or by RMS fit with the template, and appropriately incorporated into the developing molecular model. Amino acid side-chain conformations can also be extracted from known structures or from a library of standard rotamer conformations. A flexible method for specification of target geometry enables identification of substructures that conform to a variety of structural or amino acid sequence contexts. The method has been used to aid model building from crystallographic electron density maps, for homology model building, and in structure analysis applications relevant to protein engineering.

INTRODUCTION

Structural studies of proteins have shown them to be organized on the lines of a relatively limited number of structural motifs. The recognition of these motifs has generally followed an understanding of the hierarchical organization of protein structure. The secondary structural elements that predominate in fibrous proteins were understood first. Later, when more globular protein structures became known, longer range structural motifs were recognized and classified (e.g. Richardson,1981;Weber, et al.,1980; Richmond et al.,1978; Chothia et al.,1977). In practical terms these observations had relatively little impact on detailed molecular modeling tasks, such as required to build protein structures into x-ray electron density maps. Fundamentally, this reflects the relatively limited utility of knowing general features of protein structure when the task at hand involves construction of a unique and detailed molecular model that must conform to experimental data. However, the combination of interactive computer graphics, and an extensive substructure library derived from well resolved protein crystal structures, now make it possible to use detailed information about structural precedents in building models of new proteins.

Traditional implementations of computer graphics for protein modelling have concentrated on emulating the Kendrew models used originally to construct physical models of proteins. Molecular manipulations have been largely limited to the molecular degrees of freedom such as rotations about dihedral angles. Graphics programs popular with protein crystallographers, such as FRODO (Jones; 1985, 1978), have provided additional flexibility, and allow specific atoms or groups of atoms to be disconnected and relocated independent of the rest of the molecule, thereby making it easier to fit structural elements to an electron density map. Standard bond lengths, angles and torsion angles can then be restored by application of regularization procedures (Hermans et al., 1974; Jones et al., 1984). This is a powerful approach, but relies heavily on the model builder's knowledge of protein structure and conformation, since the regularization procedure cannot readily overcome barriers between local minima.

An alternative to conventional graphic modelling involves the utilization of fragments from known protein structures. Jones et al.(1986) have recently demonstrated that the entire backbone of retinol binding protein can be assembled from 22 fragments between 4 and 16 residues in length. The present work amplifies and extends this pioneering application and illustrates how the simultaneous utilization of both a backbone structural fragment and a side-chain rotamer library can substantially aid in the rapid construction of protein models from electron density maps, and additionally provide a powerful tool for structural analysis and protein engineering.

Figure 1 schematically illustrates the architecture of FRAGLE (FRAGment Locate and Exchange), a comprehensive interactive program implemented in this laboratory to examine the utility of modeling protein structures from fragments. It illustrates how commands invoked by the user direct atomic coordinate data from a library of known protein structures, through a variety of screening procedures, to eventual incorporation in a new molecular model. The design fulfills two critical requirements of a useful model building tool: 1) flexibility for use in a wide variety of applications, and 2) enough speed to allow interactive use. The following paragraphs describe specific details of the implementation.

Database Definition

In order to allow rapid and flexible search of a large number of protein structures, structural information from the Protein Data Bank (Bernstein et al.,1977) is condensed to a more readily accessible form. The residue-indexed binary file format utilized by FRODO (DSN2) is convenient for rapid retrieval of selected residue ranges, so Data Bank structures are cast in this format. Much greater efficiency in searching can be realized by precomputing distances between all $C\alpha$ atoms in each polypeptide chain, which can then be used for preliminary evaluations of chain geometry (Jones et al.,1985). The interatomic distances $d(ij)$, together with amino acid sequence information and pointers to more complete atomic coordinate data (PDB DSN2), are written into a Compacted Library of Known Protein Structures (Figure 1). To speed all comparisons to structures in the library, the distances are stored and manipulated as integers (e.g. $\mathring{A}*10$). The selection of structures to be included is determined upon the basis of structure resolution and refinement criteria. Most characterized elements of protein structure are well represented in a relatively small number of protein structures that

177

have been determined at high resolution (>1.8Å) and refined to R-values in the middle teens. For many model building applications, a library with only these few structures will suffice. Other applications, such as investigations involving structure/sequence correlation or large structural motifs, often benefit from use of a larger database. For this reason a library specification mechanism has been implemented which makes it possible to readily change from one Compacted Library to another. Since new structural data is constantly being obtained, appropriate tools for managing the library have been developed. Table 1 gives a representative list of structures included in a typical working library.

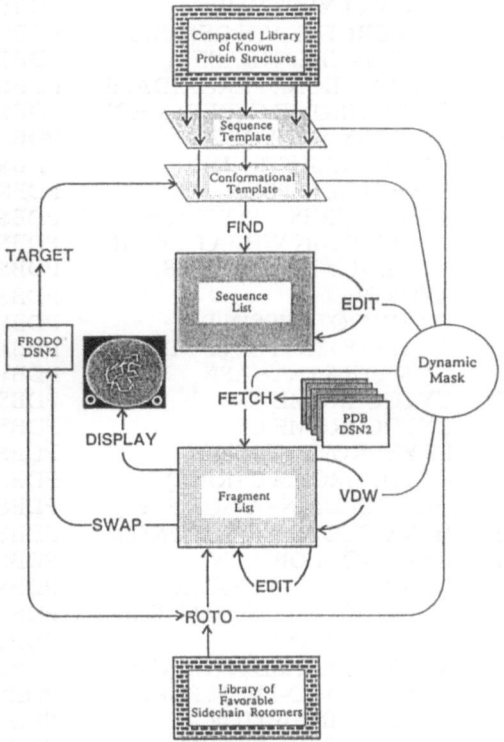

Figure 1. Schematic illustration of FRAGLE program architecture. Squares denote specific data structures; heavy lines and arrows illustrate data flow and the command under whose control the transfer is made. Other constructions denote logical conditions which influence data manipulation. The Sequence and Conformational Templates define chacteristics of protein fragments allowed to cascade from the Library of Known Structures to the Sequence List during execution of a FIND command. FETCH causes complete atomic coordinate information for fragments to be loaded from disk files (PDB DSN2) to memory (the Fragment List), from which they may be displayed, or selected to replace the original target. Sidechains conformations may be assembled from the Library of rotamers with the ROTO command. Many operations are influenced by the status of the Dynamic Mask; a logical construction to enable interactive user control of fragment manipulation (See text).

Table 1: Fragment Structure Library
(46 Structures)

ID	No. Res.	Name	PDB File
P450CAMA	405	CYTOCHROME P450 (10-250)	PDB$:NATCAM108
BPN'7113	275	SUBTILISIN BPN' (GENEX)	PDB$:SUBT80.DN2
CCP	29	CYTOCHROME C PEROXIDASE	PDB$:1CYP.DN2
CRAMBIN	46	CRAMBIN (1.5A STRUCTURE)	PDB$:1CRN.DN2
LYSOZYME	130	HUMAN LYSOZYME	PDB$:1LZ1.DN2
LC.DHFR	162	DIHYDROFOLATE REDUCTASE	PDB$:3DFR.DN2
CPEPTDAS	307	CARBOXYPEPTIDASE-A	PDB$:5CPA.DN2
G-PEROXY	185	GLUTATHIONE PEROXIDASE	PDB$:1GP1.DN2
TUNACYTC	103	CYTOCHROME C (REDUCED)	PDB$:4CYT.DN2
MYOGLOBN	153	DEOXY-MYOGLOBIN	PDB$:1MBD.DN2
HEMRTHRN	113	MET-HEMERYTRHIN	PDB$:1HMQ.DN2
PENPEPSN	323	PENICILLOPEPSIN	PDB$:2APP.DN2
SG.PRO-A	181	STREPT GRES. PROTEASE-A	PDB$:2SGA.DN2
WG AGGLU	170	AGULUTININ WHEAT GERM	PDB$:3WGA.DN2
A-L PROT	198	ALPHA-LYTIC PROTEASE	PDB$:2ALP.DN2
AZURIN	130	AZURIN	PDB$:2AZA.DN2
CHYM PI	131	A-CHYMOTRYPSIN PT.I	PDB$:5CHA.DN2
CHYM PII	97	A-CHYMOTRYPSIN PT.II	PDB$:5CHA.DN2
CIT SYN	437	CITRATE SYNTHASE	PDB$:1CTS.DN2
MOLI-CC'	127	CYTOCHROME C'	PDB$:2CCY.DN2
CC3	107	CYTOCHROME C3	PDB$:2CDV.DN2
ERYTHROC	136	ERYTHROCRUORIN	PDB$:1ECA.DN2
CYC RICE	111	CYTOCHROME C (RICE)	PDB$:1CCR.DN2
HEME A	141	HEMOGLOBIN (A-SUBUNIT)	PDB$:2HHB.DN2
HEME B	146	HEMOGLOBIN (B-SUBUNIT)	PDB$:2HHB.DN2
IMMUNO	114	IMMUNOGLOBULIN	PDB$:2RHE.DN2
LYZ T4	164	LYSOZYME (T4 PHAGE)	PDB$:2LZM.DN2
R PROT	68	L7/L12 50S RIBOSOMAL PRO	PDB$:1CTF.DN2
PLASTOC	99	PLASTOCYANIN	PDB$:5PCY.DN2
RNASE-X	124	RIBONUCLEASE-X (GENEX)	PDB$:1RSM.DN2
TRYPSIN	223	TRYP/P-AMID-PHEN-PYRU	PDB$:1TPP.DN2
TR INHIB	58	TRYPSIN INHIBITOR	PDB$:4PTI.DN2
PARAVALB	108	CARP PARVALBUMIN	PDB$:1CPV.DN2
ACD PROT	330	ACID PROTEINASE	PDB$:4APE.DN2
C ANHYD	256	CARBONIC ANHYDRASE	PDB$:2CAB.DN2
GII CRYS	174	GAMA II CRYSTALLIN	PDB$:1GCR.DN2
DHFR ECO	159	E-COLI DHFR	PDB$:4DFR.DN2
ERABUTOX	62	ERABUTOXIN B	PDB$:2EBX.DN2
FLAVODOX	138	FLAVODOXIN	PDB$:4FXN.DN2
KALKRN A	80	KALLIKREIN A (A16 A95)	PDB$:2PKA.DN2
KALKRN B	152	KALLIKREIN A (B95 B246)	PDB$:2PKA.DN2
LYZ TRIC	129	LYSOZYME TRICLINIC	PDB$:1LZT.DN2
OVO	56	OVOMUCOID 3RD DOMAIN	PDB$:2OVO.DN2
SGPB/E	185	PROTEINASE B (STREP)ENZ	PDB$:3SGB.DN2
SGPB/I	50	PROTEINASE B(STREP)INHIB	PDB$:3SGB.DN2
RMCP II	224	RAT MAST CELL PROTEINASE	PDB$:3RP2.DN2

Target Specification

The first step in a search is to define characteristics of desired structural elements. A 'target' is defined as a residue range (or ranges) of partial atomic coordinates from the current molecular model which may potentially be replaced by fragments selected from the library of known structures. The target specification establishes the length of fragments to be considered and defines (through predetermined α-Carbon positions) an approximate geometry of acceptable fragments (the Conformational Template; Figure 1). Many structural units of protein structure may not be represented by a single residue range. The adjacent strands of a twisted β-sheet, for example, have a well defined structure independent of the length of the intervening loops or the relative position of the segments in the overall amino acid sequence. The target specification allows multiple ranges to be selected to accommodate these situations. An important application of fragment fitting is the modelling of incomplete structures, where the target is not entirely defined beforehand. As outlined below, the approach followed in modeling such regions will differ depending on the target specification required in a particular application.

Masking

In order to improve the adaptability of the target specification toward accommodating the specific needs of the user, two logical constructions are defined that collectively constitute a 'Dynamic Mask' (Figure 1). The first is a vector (a(n)) of logical quantities (where n is the number of residues in the target) that flag

Target Sequence	[YNQLSGTF]
Mask Status	[---------S-----l]
	22 29

Sequence Selection of Specific Amino Acid by Class:

Key	Class	Allowed Amino Acid
' '	Any	ACDEFGHIKLMPQRSTVWY
'b'	Beta	CFOVWY
'h'	Helical	AEHKLMQR
't'	Turn	DGNPST
'l'	Large	FHMWY
's'	Small	AGS
'y'	CB Branched	ITV
'n'	Nonpolar	ACFILMVWY
'p'	Polar	DEHKN QRST
'+'	Positive	HKR
'-'	Negative	DE

Figure 2. Dynamic Mask Sequence Designation. The library of x-ray structures may be searched using a flexible amino acid sequence and property mask. The mask illustrated confines the search to fragments whose backbone α–Carbons fit the target to some prespecified tolerance, and which also incorporate a serine residue in the position corresponding to residue 26 of the target and a "large" residue at position 29.

whether or not a particular residue is 'active' or 'inactive' in the context of specific operations defined below. The second is a matrix (20 by n) of logical quantities which flag the activity or inactivity of each of the twenty amino acids at each target residue position. The latter mask is used primarily as a means of specifying amino acid sequence or residue type (eg. polar, charged, etc.) requirements for a given fragment match to a target. The status of the Dynamic Mask is constantly indicated in a menu-like display and can be adjusted at any time (Figure 2).

Searching The Database of Known Structures

All potential protein fragments from the library of known structures are screened for compatibility with the input target. Incompatible fragments are eliminated as efficiently as possible. A potential fragment is first tested against the sequence requirements given by the user in the specification of the Dynamic Mask, where any discrepancy between target and fragment sequences causes immediate rejection.

To screen out fragments of inappropriate geometry, individual elements of a triangular matrix of inter-$C\alpha$ distances characteristic of the target $(dt(i,j))$ are compared with corresponding elements precomputed for the potential fragment $(df(i,j))$. Any difference $|dt(i,j)-df(i,j)|$ greater than a user selected tolerance (e.g. 1.0 Å) causes rejection of the fragment. The search through all library structures can be made efficient by comparing interatomic distances $d(1,n)$ first, then backwards to $d(1,2)$, since distances between α-Carbon atoms show more variability as the number of residues between them increases. Distance correspondence between $C\alpha$'s is only required at active residue positions, as defined by the status of the Dynamic Mask. Any distance $d(i,j)$, where i or j is an index to an inactive sequence position is ignored. This makes it possible to search for a fragment loop with particular endpoint geometry, while making no requirements on the intervening structure except the number of residues.

For targets involving more than one sequentially connected residue range, the above procedure is first used to find fragments compatible with the first segment. The characteristic relationships between these residues and residues of other target segments are then investigated to identify additional fragment segments over the entire length of the polypeptide.

The result of a search is a list of sequences which comply within the requested tolerance. Since the gathering and superposition of complete atomic coordinate data is relatively slow, it is useful to be able to rank these sequences prior to subsequent processing. For each fragment recovered, a mean square deviation $(\Delta[d(i,j)])$ from the target is obtained during comparison of interatomic distance matrices, and a numerical sequence homology score (Dayhoff et al., 1978) derived from comparison of the target and fragment amino acid content. An editing facility has been implemented to enable sorting of the sequence list using either of these criteria or selection of specific entries for preservation or deletion.

Superposition

The superposition of the target and fragments is accomplished by the method of (Kabsch ,1978) based on a list of atom correspondences. Since the RMS difference in position of individual atoms after transformation is often the most quantitative gauge of the quality of the fit of a fragment to the target, it is important to be able to adjust which atoms will be included in this mean. Whole residues can be excluded from a correspondence list by inactivating the residue position in the Dynamic Mask. Further specificity can be achieved to defining a set of atom names (e.g. N CA C O) that will be forced to correspond. Unnamed atoms are ignored.

Fragment Disposition

Once fragments have been identified and oriented coincident with the target, fragments may be displayed for evaluation on the graphics screen, or subjected to screening under van der Waal's restraints. The user may eventually select a fragment to replace complete atomic coordinate information of the target. As in the case of superposition, precise control over the disposition of each atom is possible during the replacement process.

Side-Chain Modeling

Surveys examining the occurence of amino acid side-chain conformations in known protein crystal structures have demonstrated a preference for a limited number of low energy conformations (Janin et al., 1978; Bhat et al., 1979; James et al., 1983). A more recent survey of highly refined crystal structures (Ponder et al., 1987) has confirmed the earlier findings, and concluded that deviations from ideal stereochemistry are more infrequent than previously thought. We have implemented a procedure which allow all common conformations of a side-chain to be placed on a backbone. (The backbone atomic positions must be already defined). Side-chains are extracted from a library of favorable rotamers (Ponder et al.; 1987) and placed upon the target backbone. Rotamers may be manipulated exactly as fragments extracted from known structures, displayed for evaluation on the graphics system and selected to replace the target model.

Implementation

The fragment manipulation capabilities of FRAGLE have been implemented as a module within FRODO (Jones, 1985). In this way, all the functionalities (such as electron density display, model manipulation and geometry regularization) indispensable to crystallographic applications are retained. Our implementation utilizes a version of FRODO specific to Evans and Sutherland PS300 series graphics devices (Pflugrath et al.,1984), although it may be easily modified for any hardware to which FRODO has been adapted.

APPLICATIONS

Electron Density Map Fitting

FRAGLE can be used very effectively, in conjunction with model building capabilities incorporated in FRODO, as an aid in electron density map fitting. A recent application of FRAGLE in molecular model construction was the structure determination of protocatechuate 3,4 dioxygenase (EC 1.13.1.3) (PCDase), an enzyme from *Pseudomonas aeruginosa* that catalyses the the oxygenolytic cleavage of protocatechuic acid (3,4 dihydroxybenzoic acid) into β-carboxy-cis,cis-muconic acid. The holoenzyme is a 587000 dalton dodecamer of protamers arranged with 23 local symmetry. Each protamer is composed of 2 polypeptide chains containing a total of 440 amino acids.

The structure of PCD was solved using a combination of multiple isomorphous replacement and non-crystallographic symmetry averaging methods to produce a final, 6-fold symmetry averaged, map at 2.8 Å resolution (Ohlendorf et al., 1988). This map was completely interpreted without the use of a minimap using FRODO/FRAGLE on an Evans and Sutherland PS330 graphics system. The procedure started by positioning a string of points representing α-Carbons positions separated by 3.8 Å throughout the electron density map. The string was oriented so that the α-Carbons 1 and 2 were in their proper locations in the density. The bond between α-Carbons 1 and 2 was then broken, and the string rotated about α-Carbon 2 to position α-Carbon 3. This process was repeated to determine successive α-Carbon atom positions until the density ended or became ambiguous. In this manner over 90 percent of the α-Carbons were placed in the map.

In the next stage, the α-Carbons were replaced with complete polypeptide backbone by searching the library of refined protein structures (Table 1) for α-Carbon segments which gave the best RMS fits to the α-Carbon string originally built into the electron density map. Usually this search involved 5 to 9 residue segments. The best fitting structures were visually inspected and the one which optimally fit β-Carbons and carbonyl oxygens into the map density was incorporated into the model. In adding successive segments to the growing model, the terminal residues were generally not incorporated, since the ends had a tendency to fray due to lack of constraints on the course of chain. Finally, side chains were positioned in the density and added to the structure. FRAGLE displayed the most common rotamers for each residue as defined from the rotamer library (Ponder et al., 1987). Generally, a single rotamer conformation could be uniquely selected that best fit the electron density and was subsequently incorporated into the model.

PCDase contains large amounts of regular secondary structure, primarily in the form of β sheet. In these regions relatively long fragments (up to 13 residues) could be built as a single segment. Turns generally required shorter segments and often fit the density less well. Since turns generally have higher temperature factors, their densities are generally weaker. Also, some turns were not well represented in the data base. Nevertheless, it was possible using FRAGLE to routinely build 40 residues per day. The process is outlined in Figure 3.

In addition to speed in model construction, the use of the structural data base incorporated in FRAGLE provides important conformational information that may not be readily evident from initial inspection of the electron density map. For example, if a β strand is being built, and the 12 best examples all have a carbonyl group in a particular position, one is willing to accept it even though the map might be ambiguous about its placement. Similar situations also occur frequently with side chain placements.

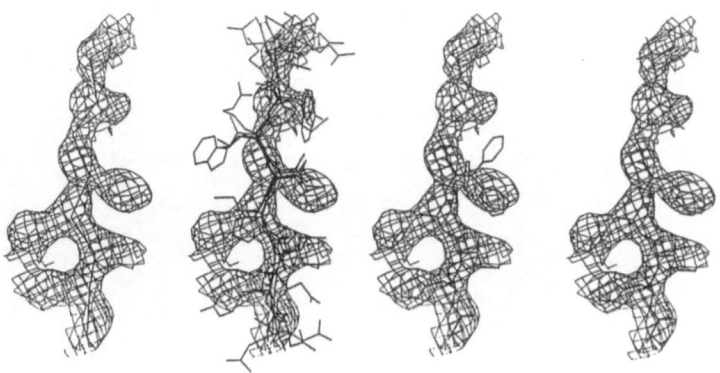

Figure 3. The panels illustrate the isolated electron density for a β strand in the 2.8 Å resolution x-ray map of PCDase. Successive panels show map with 3.8 Å αC string, superimposed polypeptide backbone structures retrieved from the database, the selected polypeptide backbone with multiple side chain rotamers displayed, and the final model structure.

As a result of incorporating fundamentally correct stereochemistry during model building, the final structures produced with FRAGLE have excellent geometry, a result which saves considerable subsequent effort in crystallographic refinement. In the case of PCDase, the model built with FRAGLE produced an initial R value of 0.38 for for 20,616 atoms and 59,057 data to 3.0 Å resolution (Ohlendorf et al., 1988). Additional examples of the use of FRAGLE to build crystal structures can be found in Weber et al., (1989).

Molecular Analysis

Comparison of a collection of protein fragments with similar polypeptide conformation frequently reveals other structural characteristics common to the ensemble which had gone unrecognized upon examination of structures individually. Often these characteristics are simple or predictable (such as the requirement for a specific amino acid at a position in a tight turn, or the preference for a particular conformation of threonine side-chains in an extended α-helix), but more complicated generalities may also be revealed. In many cases, these localized motifs provide important information potentially relevent to engineering protein structures (Richardson et al., 1988; Blundell et al., 1987). FRAGLE has been extensively used to search for or verify such patterns, two examples of which are given below and illustrated in Figure 4.

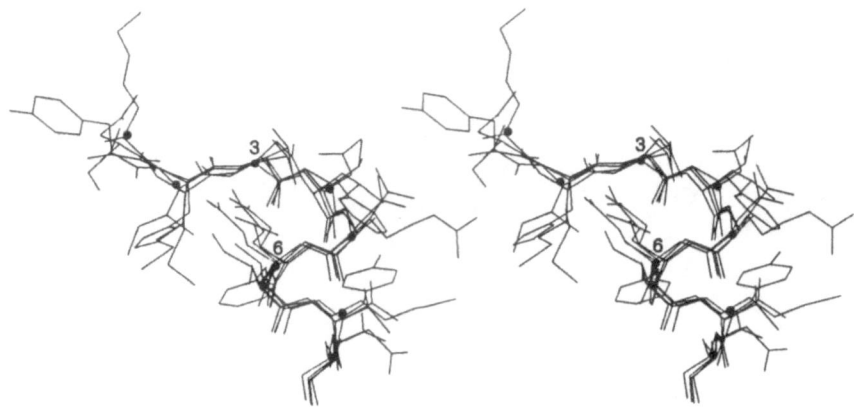

Figure 4. Stereoscopic representation of four similar protein fragments extracted from the library of known structures. The four (and the amino acid sequence) are from 1) Cytochrome P450 residues 190-198 (SMTFAEAKE); 2) Cytochrome c Peroxidase residues 162-170 (NMNDREVVA); 3) Carboxypeptidase residues 12-20 (YHTLDEIYD); and 4) Parvalbumin residues 96-104 (KIGVDEFTA). All are superimposed on the C-α backbone of Calmodulin residues 99-107 (FISAAELRH) (shown as dots) used as a target conformational template.

In Figure 4, four structurally homologous protein fragments are displayed. The four were selected as the only examples from a library of 40 known protein structures which meet a stringent (0.7 Å) tolerance in comparison to a nine residue section of the calmodulin α-Carbon backbone (Babu et. al., 1985). The polypeptide conformation adopted by all these fragments consists of an extended β conformation that loops immediately into an α helix. Results of the search demonstrate a preference for short oxygen-containing side-chains (serine, threonine or asparagine) at the third residue position, where a hydrogen bond can be accepted from the amide at position 6. Examination of the fragment sequences also reveals a strong preference for glutamic acid at position 6 in helical N terminii of this geometry, possibly because of potential interactions between the side-chain carboxylate and the amide of residue 3. The existence of these specific interactions in a variety of otherwise unrelated protein structures illustrates the importance of these "capping" interactions in the stabilization of this conformation, and a possible mechanism by which both serine and glutamic acid contribute to the initiation of helix formation in globular proteins (Robson et al.,1972, Richardson et al., 1988).

Another example typifies the sort of fragment analysis which has potential utility in prediction of amino acid side-chain conformation. Surveys of side-chain conformations, such as that of (Ponder et al., 1987), have shown higher frequency of occurrence for some conformations than others. However, the conformation adopted at a particular position in a protein structure is highly dependent upon the local conformation of the polypeptide chain and other contextual factors. Serine, for example, can adopt three common side-chain conformations: + (chi=+60°), - (chi=-60°), and t (chi=180°), of which the + conformation occurs most frequently. For fragments in an extended β conformation, the - conformation occurs with higher frequency (41%, vs. 28% + and 32% t). If only fragments from extended parallel β sheets are selected (Figure 5), it can be shown that only the - conformation is observed, presumably because of packing restrictions imposed by the close proximity of neighboring strands. The same conclusion cannot be drawn for serine in anti-parallel β sheets, where other serine side-chain conformations are frequently observed. While this is a very specific example, analyses of side-chain conformation frequencies among fragments of homologous structure can often provide convincing evidence to support the selection of one conformation over others.

Structure Generation From Backbone Coordinates and Homology Building

In many cases, it is necessary or desirable to generate an essentially complete molecular model when only α-Carbon coordinates may be available in a database. Alternatively, it may be that coordinates are available for one species of a protein, but not those of a related, homologous species. In both cases it is possible to use the fragment replacement capabilities of FRAGLE to extend the input structure.

In one such study (Weber et. al., 1990), the objective was to study the consequences of alterations in surface charge residues on the functional properties of plant calmodulin mutants that were highly homologous to the vertebrate protein. When this study was begun, only α–Carbon coordinates of the protein were available, so that it was necessary to reconstruct the entire structure from these partial structural data. This case was felt to be particularly favorable for two reasons: 1) Calmodulin is composed predominantly of α helices, fragments which are very well represented in the structural database. 2) The objective involved the computation of molecular electrostatic fields, which are relatively insensitive to small errors in side-chain placement (compared, say, to the precision required to accurately reconstruct an enzyme active site.)

In this example, α-Carbon coordinates from rat testis calmodulin (Babu et al, 1985) were extended using the fragment fitting strategies outlined above to produce a complete polypeptide backbone structure. Indeed, experience has now shown that location of α–Carbon positions is nearly always sufficient to correctly orient the polypeptide backbone planes within a few tens of degrees. Since this also correctly orients the Cα-Cβ bond, the only remaining variables to be determined involve selection of the proper side chain rotamers. As illustrated above and described also in (McGregor et al., 1987), side chain rotamers can in many cases be predicted owing to their incorporation in a particular secondary structure. Thus, side chains were added to calmodulin by selecting 3 to 5 residue fragments whose backbone matched the input α-Carbon target, using a mask that required amino acid identity in the side chain position to be substituted. Unfavorable steric interactions that occurred in the building process were relieved either by selection of an alternative allowed side chain rotamer, or by energy minimization. This process produces a very well packed protein structure with solvent exposed charge groups and virtually all essential features of the

186

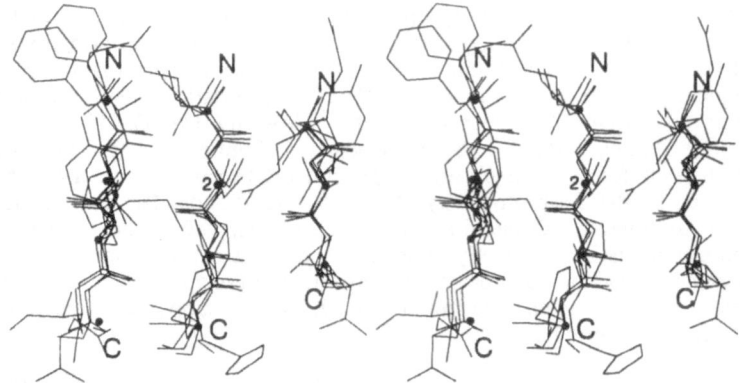

Figure 5. Stereoscopic illustration of the conformation of all occurences of three contiguous parallel β strands from the library of known structures which have a serine at amino acid position 2 of the central strand. Selected from four different protein structures, all have the same conformation for the serine sidechain. The target geometry was specified using α-Carbon positions from Subtilisin BPN' (residues 28-30, 121-124 and 148-151), which utilizes an isoleucine at this position (residue 122).

calcium binding site of the related protein parvalbumin (not included in the structure library used for this study) properly regenerated. These mutants proteins are now being examined crystallographically, and it will be of some interest to examine the accuracy of the prediction.

References

Babu, Y.S., Sack, J.S., Greenhough, T.J., Bugg, C.E., Means, A.R. and Cook, W. J., Three-dimensional Structure of Calmodulin. Nature, **315**, 37-40 (1985).

Bernstein, F.C., Koetzle, T.F., Williams, G.J.B., Meyer, E.F.Jr., Brice, M.D., Rodgers, J.R., Kennard, O., Shimanouchi, T. and Tasumi, M., The Protein Data Bank: A Computer-based Archival File for Macromolecular Structures. J. Mol. Biol., **112**, 535-542 (1977).

Bhat, T. N., Sasisekharan, V. and Vijayan, M., An Analysis of Side-chain Conformations in Protein Structures. Int. J. Pept. Res.,**13**, 170-184 (1979).

Blundell, T. L., Sibanda, B. L., Sternberg, M.J.E. & Thornton, J. M., Knowledge-based Prediction of Protein Structures and the Design of Novel Molecules. Nature, **26**, 347-352 (1987).

Chothia, C., Levitt, M., and Richardson, D. Structure of Proteins: Packing of α-Helices and Pleated Sheets, Proc. Natl. Acad. Sci. USA **74**, 4130-4134 (1977).

Dayhoff, M. O., Schwartz, R. M., and Orcutt, B. C., A Model of Evolutionary Change in Protein Structures. Atlas of Protein Sequence and Structure. (1978)Vol. 5 Suppl.3 (M. O. Dayhoff, ed.). pp. 345-352 (1987).

Hermans, J. & McQueen, J.E., Computer Manipulation of Macromolecules with-the Method of Local Change. Acta Crystallogr., A30, 730 (1974).

James, M.N.G. and Sielecki, A.R. Structure and Refinement of Penicillopepsin at 1.8 Å Resolution J. Mol. Biol., 183, 299-361 (1983) .

Janin, J., Wodak, S., Levitt, M. and Maigret, B. Conformation of Amino Acid Side-chains in Proteins. J. Mol. Biol., 125, 357-386 (1978).

Jones, T.A., Interactive Computer Graphics: FRODO Meth. Enzymol., 115,157-171 (1985).

Jones, T.A. & Liljas, L., Crystallographic Refinement of Macromolecules having Non-crystallographic Symmetry. Acta Crystallogr., A40, 50 (1984).

Jones, T.A. & Thirup, S., Using Known Substructures in Protein Model Building and Crystallography EMBO J., 5, 819-822 (1986).

Kabsch, W., A Discussion of the Solution for the Best Rotation to Relate Two Sets of Vectors. Acta Crystallogr., A34, 827-828 (1978).

McGregor, M., Islam, S. A., and Sternberg, M. J. E., Analysis of the Relationship Between Side-chain Conformation and Secondary Structure in Globular Proteins. J. Mol. Biol. 198, 295-310 (1987).

Pflugrath, J.W., Saper, M.A. and Quiocho, F.A. (1984) in "Methods and Applications in Crystallographic Computing" (S. Hall & T. Ashida, eds.) Oxford Univ. Press, London, pp 404.

Ponder, J.W. & Richards, F.M., Tertiary Templates for Proteins. J. Mol. Biol., 193, 775-791 (1987).

Richardson, J.S. The Anatomy and Taxonomy of Protein Structure. Adv. in Protein Chem., 134, 167-338 (1981).

Richardson, J.S., and Richardson, D.C., Amino Acid Preferences for Specific Locations at the Ends of α-Helices. Science, 240, 1648-1652 (1988).

Richmond, T.J and Richards, F.M. Packing of α-Helices: Geometrical Constraints and Contact Areas, J. Mol. Biol. 119, 537-555 (1978).

Robson, B. and Pain, R.H., Directional Information Transfer in Protein Helices. Nature, 238, 107-108 (1972).

Weber, P. C.,Lukas, T. J., Craig, T. A., Wilson, E,. King, M. M., Kwiatkowski, A. P. and Watterson, D. M., Computational and Site-Specific Mutagenesis

Analyses of the Asymmetric Charge Distribution on Calmodulin. (in press) PROTEINS:Structure, Function and Genetics (1990).

Weber, P. C. and Salemme, F. R., Structural and Functional Diversity in 4-α-Helical Proteins. Nature **5777,** 82-84 (1980).

Weber, P. C., Ohlendorf, D.H., Wendoloski, J. J., and Salemme, F.R. Structural Origins of High-Affinity Biotin Binding to Streptavidin. Science **243,** 85-88 (1989).

O: A Macromolecule Modeling Environment

T. Alwyn Jones, Marc Bergdoll, and Morten Kjeldgaard

Aims

The last ten years has shown a tremendous increased interest in structural molecular biology. This has been partly fuelled by the revolution in molecular biology whereby it is now possible to express large amounts of material for proteins that normally exist in very small quantities in the cell. There have also been important developments in crystallography and NMR. Computer graphics has played a small part in these developments, allowing the scientist to build, manipulate and interact with his molecule (see Jones, 1987, for a review).

The program Frodo (Jones, 1978, 1982) has been widely used on many different computer and displays. It has been modified to make use of technological developments (e.g. the use of colour) and to enhance functionality (Jones & Liljas, 1984, Pflugrath et al., 1984, Jones & Thirup, 1986). Although it was possible to implement data base modelling in Frodo (Jones & Thirup, 1986), it was clearly time to design a new program. This new program should use a general data base, allow for parallel developments in a number of laboratories, be easy to use and should be easily transferred between different computers and work stations.

The first release of this program (called O) is now available. It is already our program of choice for building a new structure in an electron density map. Our belief in using a data base has been validated my ease of both working with the program and developing new options. This release lacks some Frodo functionality (in particular the abilities to mutate the protein), but this situation will be corrected. At present the program runs on a DEC Vax/Evans & Sutherland PS300. A Stellar GS1000 PHIGS/X-windows based version is nearly completed.

O Data base

This consists of vectors of data types that we call parameter blocks. These data types may be integer, real, character variables and text. The vectors are named and have various states. The most important states define read/write access and whether the data should be deleted. Most data used or created by the program sit in the data base. Coordinates, for example, exist as a series of vectors with a naming convention so that a file used by the Prolsq program (Hendrickson & Konnert, 1980) is decomposed into a minimum of 7 vectors (see listings from a simple interactive session in appendix 1). In this naming convention characters 1-5 define the molecule name; if characters 6-14 are '_RESIDUE_' then the vector contains residue related data (we call these residue properties), or if 6-11 are '_ATOM_' then the vector contains atom related data (so called atom properties). Other atom or residue properties can be generated by the user (on line or by using an editor) or by using separate stand-alone programs. For example, secondary structure residue properties are generated by a slightly modified version of the DSSP program (Kabsch & Sanders, 1983).

The forcing of a naming convention is a general feature of the program and may be transparent to the casual user. For example, the molecule comparison features store four vectors in the data base for each comparison saved by the user. These entries describe the transformation operator, the names of the two molecules compared, the atoms used and an graphics object description showing coloured vectors between paired atoms (after applying the transformation to one molecule). The user gives a character sting for his comparison experiment to which the program forces its own convention. For example if the result of comparing BBP to RBP is called 'BBP_TO_RBP', then the rotation translation operator is actually stored in '.LSQ_RT_BBP_TO_RBP'.

Another useful group of items kept in the data base are data that are default values that do not need to be often changed. For example, the object definition code for how to treat a single, unconnected atom is in the data base. This is usually a 3-D cross but could be changed to a tetrahedron, icosahedron, sphere etc. Almost every major menu has some piece of data stored in the data base.

Gone are the days of complaining about the names in a menu, these are a part of the data base and can be changed.

An example of how a user may use the data base is given in the macro discussed in appendix 2.

Menus

Each major piece of functionality exists as a grouping of keywords. Such a major menu can expand to a maximum of ten minor keywords. This is intended to make it easier for both developers and users. Flag codes are issued to developers ten at a time without caring if they are all used. The user can decide which major menus can be activated and whether they are to be activated by picking at the terminal and/or by typing. The command line interpreter allows any number of commands per line, part typing of keywords (e.g. 's_a_i' is enough to activate 'sam_atoms_in'), prompting for every not supplied value, and extensive defaults (see Appendix 1).

Beginners are taught to limit and personalise the keywords appearing on the screen. Since the program can use macro files of instructions which can also be activated by picking, we encourage a simple screen layout of one major menu of frequently used commands (centre, yes, no etc.), two menus of personalised macros, and one menu controlling object visibility.

Developers keep to a set of programming rules to ensure lack of interference between major menus. Parallel development work is therefore painless.

The current program has twenty major menus, of which four are devoted to user macros and object control. The following briefly describes some of the remaining 16 menus.

_Draw_Mol

The data base can hold any number of molecules. In itself it contains no limitation of what is in a molecule, how many chains there are etc. The program can display any number of molecular objects. An object is associated with only molecule, but a molecule can be used to create many different objects.

A molecular object is created by combining molecule drawing commands to display a zone of atoms, Cα trace, residues within a sphere of a point or residues covering an atom or residue. Bonds appear from a connectivity file which can be changed during object creation, so that a tRNA phosphate backbone can easily be drawn with a Cα backbone. The sphere command is centred on a 3-D point in the data base named '.active_centre'. This is a coordinate like any other and can be modified by arithmetic operations (see the macro in appendix 2).

_Map

The map data structure is the same as in Frodo (Jones & Liljas, 1984) and is therefore not part of the data base. There is a one to one correspondence between

192

a single contour level and an object. Since any number of objects can be drawn, there is no limitation to the number of levels, or maps, or files.

_RSR

The same map structure is used for real space refinement. This refers to a technique whereby a model is changed to optimize its fit to the observed electron density (Diamond, 1971; Jones & Liljas, 1984). This is a greatly extended implementation compared to Frodo. It includes residue library fragment fitting (Jones & Liljas, 1984), main-chain, rotamer, and dihedral angle fitting.

_Bones

The use of skeletonised electron density with a protein data base (Jones & Thirup, 1986) has greatly improved the initial stage of map interpretation and model building. The skeleton is produced with the algorithm of Greer (1974). It is first used to test hypotheses to determine the polypeptide chain fold and then, in combination with a protein data base, to act as a framework on which to build the backbone. In Uppsala we have solved a variety of structures with initial maps of varying quality. These include rubisco from Rh. rubrum containing 950 residues in the asymmetric unit, 2.9 Å resolution (Schneider et al., 1986), spinach rubisco (2400 residues, 2.8 Å, Andersson et al., 1989), P2 myelin (400 residues, 2.7 Å, Jones et al., 1988), T. reesei cellulase CBH2 (750 residues, 2.8 Å, Rouvinen et al., in preparation, 1989), E. coli ribonucleotide reductase B2 (750 residues, 2.7 Å. Eklund et al., in preparation, 1989).

Each skeletonised map becomes a molecule in the data base with two extra entries, one defining its connectivity, the other an atomic property used to define if the bone is considered a main chain, side chain or anything else. As for all molecules, an object is made from just a single skeleton molecule. Any number of objects can be drawn making it convenient to display different skeletons, with different properties, with different radii and different colours. The usual practise for map interpretation, therefore, is to make one object with a large selection radius showing the current main chain interpretation, together with a smaller radius object showing all of the skeleton atoms. The properties and connectivity of the atoms in the latter object would then be changed according to the users interpretation of the electron density.

The connectivity can be changed by commands to make new bonds and break existing linked bonds.

The assignment of guide Cα positions on the skeleton is different from Frodo. In O the user directly picks which bones atom coordinate is to be assigned to a particular Cα in the sequence. The user is shown a window into the sequence which can be moved forwards or backwards. The coordinates of the identified bones atoms then gets used for the Cα of the residue in the centre of the window.

_Proleg

This menu allows proteins to be treated like pieces of Lego. This is a two stage process, first to build the main chain and then to add on the side chain. Two data bases are available. One data base fits Cα or skeleton guide points to the main chain of a library of well refined proteins (Jones & Thirup, 1986), the other uses a rotamer side chain data base (Ponder & Richards, 1987) to build the side chain. An autobuild command constructs a molecule from Cα atoms.

_Refi

The Hermans & McQueen (1974) model regularisation has been implemented.

_Manip

In our first implementation, we provide only a single atom move. This was to force the use of our data base building options on the initial users. We are now expanding the menu to include various fragment manipulation commands.

_Lsq_align

One frequently wishes to compare different but structurally related molecules. This menu allows one to explicitly define what atoms to match between molecules A & B. The results are stored in the data base and the superposition can be made on the screen. A set of vectors can be drawn between paired atoms and is useful to see separate domain or motif movements.

The initial, often very approximate transformation can be improved. Like other algorihms (Rossmann & Argos, 1975, Rossmann & Argos, 1976), this balances goodness of fit against number of connected residues. After applying the initial transformation, the longest fragment is found such that each matched pair in the fragment is separated by less than a predefined value. The next longest fragment is then located and so on until a minimum fragment size is obtained. The paired atoms are then used to make a new transformation matrix (Kabsch 1978). The process is repeated until either the number of atoms paired off is constant or until 10 cycles are made. The pairing distance and the minimum fragment size are in the data base with default values of 3.8 Å and 4 residues. The results of the refinement can be placed in the data base and can be used as described in the last paragraph.

Appendix 1 shows an alignment of two non-crystallographically related P2 myelin models. The initial alignment is deliberately 5 residues out of phase, but immediately converges. The convergence properties of the algorithm are very good but can lead to an incorrect solution. For example, if one helix in A is matched to a helix in B such that the alignment is out by one or two residues, then it may arrive at a solution where just the helices fit.

_Paint_Mol

Molecules can be coloured by logically combining atom and residue properties (e.g. one can paint red all tryptophan residues that have accessibility greater then 20 Å^2 and are on a helix). One can also use colour ramping and by property value.

_Sketch_Mol

This allows one to view cylinder and ribbon representations of secondary structure as well as spheres and ball and stick models. The secondary structures are normally drawn based on secondary structure residue properties calculated with DSSP.

O is a useful program for inspecting/building macromolecules. We intend to make it more so by improving our existing algorithms and implementing new ideas. In particular we wish to make more use of the powerful computers available in modern work stations.

References

Andersson, I. et al. (1989). Nature 337:6204, 229-234.

Diamond, R. (1971). Acta Cryst. A27, 436-452.

Greer, J. (1974). J. Mol. Biol. 82, 279-301.

Hendrickson W.A. & Konnert, J. (1980). Computing in Crystallography, 13.01-13.25, Indian Academy of Science.

Hermans, J. & McQueen, J.E. (1974), Acta Cryst, A30, 730-739.

Jones, T.A. (1978). J. Appl. Cryst 11, 268-272.

Jones, T.A. (1982). Computional Crystallography (Ed. D. Sayre), pp. 303-317. Clarendon Press, Oxford.

Jones, T.A. (1987). Crystallography in Molecular Biology. Edited by D. Moras, J. Drenth, B. Strandberg, D. Suck & K. Wilson, pp. 125-130. Plenum Press.

Jones, T.A. & Liljas, L. (1984). Acta Cryst A40, 50-57.

Jones, T.A. & Thirup, S. (1986) EMBO J. 5, 819-822.

Jones, T. A. et al. (1988). EMBO J. 7:6, 1597-1604.

Kabsch, W. (1978). Acta Cryst. A34, 827-828.

Kabsch, W. & Sanders, C. (1983). Biopolymers 22, 2577-2637.

Pflugrath, J.W. et al. (1984) In Method and Applications in Crystallographic Computing (Hall. S. & Ashida., eds.), pp. 404-407. Clarendon Press, Oxford.

Ponder, J.W. & Richards, F.M. (1987). J. Mol. Biol. 193, 775-791.

Rossmann, M.G. & Argos, P. (1975). J. Biol. Chem. 250:18, 7525-7532.

Rossmann, M.G. & Argos, P. (1976). J. Mol. Biol. 105, 75-95.

Schneider, G. et al. (1986). J. Mol. Biol. 187, 141-183.

Appendix 1.   A simple session

The following demonstrates some of the commands available to the user in O and
the nature of the terminal interface. The files defined in the beginning are data
base files with the necessary start-up data, defining such things as the menu, the
display etc. The two coordinate sets are from the P2 myelin structure.

```
$ r oexe
Echo and Loggings are set OFF.
 O > O version 4 in preparation, Feb 89
 O > Define an O file (terminate with blank): odat:startup.o
 O > File is formatted
 O > Define an O file (terminate with blank): odat:ps390.o
 O > File is formatted
 O > Define an O file (terminate with blank):
 O > File_display_connectivity is not defined.
 O > Enter file name [ odat:o.dat] :
Maximum inter-residue link distance = 6.00
There were     22 residues
               113 atoms
 O > Do you want to yse the display? [Y]/N
 O > ! read in coords for 2 chains, call them A and B
 O > sam_at_in
Sam>Name of input file: m17_a.wah
Sam>O associated molecule name : a
Sam>Type of coordinates assumed from file name.
Sam>Is that O.K. ? ([Y]/N)
Sam>Is it Wayne s format ([Y]/N) ?
Sam>Are there any S-S bridges (Y/[N]) ?
 O > sam_at_in
Sam>Name of input file: m17_b.wah
Sam>O associated molecule name : b
Sam>Type of coordinates assumed from file name.
Sam>Is that O.K. ? ([Y]/N)
Sam>Is it Wayne s format ([Y]/N) ?
Sam>Are there any S-S bridges (Y/[N]) ?
 O > dir a*
Heap>A     _ATOM_XYZ                R W 3162
Heap>A     _ATOM_B_WT_NAME_TYPE I W 1054
Heap>A     _ATOMS_IN_RESIDUES   C W   51
Heap>A     _RESIDUE_NAME            C W  132
Heap>A     _RESIDUE_TYPE            C W  132
Heap>A     _RESIDUE_POINTERS    I W  264
Heap>A     _RESIDUE_CG              R W  528

 O > save
As1> File_O_save is not defined.
As1> Enter file name [ binary.o] :
 O > ! make an object from molecule a
 O > mol
 O > Current molecule      has not been loaded.
Mol> Molecule code name []: a
 O > ca
```

```
Mol> Ca zone [all molecule]:
 O > end
 O > ! that could all have been done in one line
 O > mol a ca ; end
 O > ! centre it on the fatty acid
 O > cen_atom
As3> Define molecule [A      ], residue, and atom [CA] : a132 c5
 O > ! believe me , it did centre
 O > ! make an object with the fatty acid too
 O > ca ; zon a132 ; end
 O > ! note the use of ';' in the above line
 O > ! make another object with those residues close to the FA
 O > obj sph_a cover a132 ; 2.5 end
 O > ! after activating 'obj_on/off' I can click the objects on/off
 O > ! make mol b
 O > mol b ca ; z b132 end
Mol> Second residue not in molecule
 O > mol b ca ; z b132 ; end ! notice my error in the last line
Mol> Cannot change molecule once started
 O > end mol b ca ; z b132 ; end ! notice my error in the last line
 O > ! now compare molecules A and B
 O > lsq_ex
Lsq >Lsq definition defaults are taken.
Lsq >Least squares match by explicit definition of atoms.
Lsq >Given 2 molecules A,B the transformation rotates B onto A
Lsq >What is the name of molecule A ? a
Lsq >What is the name of molecule B ? b
Lsq >Now define what atoms in A are to be matched to B.
Lsq >Defining 3 names implies a zone and an atom name.
Lsq >Defining 2 names implies a zone and CA atoms.
Lsq >Defining 1 name  implies the CA of that residue.
Lsq >A blank line terminates input.
Lsq >Define atoms from molecule A : a1 a131
Lsq >Define atoms from molecule B : b1 b131
Lsq >Define atoms from molecule A :
Lsq >The    131 atoms have an r.m.s. fit of     0.614
Lsq >xyz(1) =     0.3866*x+    0.0907*y+    0.9178*z+  -16.9487
Lsq >xyz(2) =     0.1155*x+    0.9826*y+   -0.1458*z+   25.7526
Lsq >xyz(3) =    -0.9150*x+    0.1624*y+    0.3694*z+   66.9966
Lsq >The transformation can be stored in O.
Lsq >A blank is taken to mean do not store anything
Lsq >The transformation will be stored in .LSQ_RT_ b_to_a
 O > app_obj
Lsq >Apply a transformation to an existing object.
Lsq >There is an alignment called B_TO_A
Lsq >Which alignment [<CR>=restore a transformed object] ? b_to_a
Lsq >There is an object called A
Lsq >There is an object called SPH_A
Lsq >There is an object called B
Lsq >Which object ? b
 O > ! now they are on top of each other
 O > !I can identify them
 O > app_obj ; b
 O > ! that removed them
 O > ! do a deliberate mistake in LSQ
 O > lsq_e
```

```
Lsq >Least squares match by explicit definition of atoms.
Lsq >Given 2 molecules A,B the transformation rotates B onto A
Lsq >What is the name of molecule A ? a
Lsq >What is the name of molecule B ? b
Lsq >Now define what atoms in A are to be matched to B.
Lsq >Defining 3 names implies a zone and an atom name.
Lsq >Defining 2 names implies a zone and CA atoms.
Lsq >Defining 1 name  implies the CA of that residue.
Lsq >A blank line terminates input.
Lsq >Define atoms from molecule A : a1 a125
Lsq >Define atoms from molecule B : b5 b129
Lsq >Define atoms from molecule A :
Lsq >The    125 atoms have an r.m.s. fit of    11.038
Lsq >xyz(1) =     0.2809*x+    0.0037*y+    0.9597*z+    -8.7543
Lsq >xyz(2) =     0.2134*x+    0.9747*y+   -0.0662*z+    16.6426
Lsq >xyz(3) =    -0.9357*x+    0.2234*y+    0.2730*z+    69.5881
Lsq >The transformation can be stored in O.
Lsq >A blank is taken to mean do not store anything
Lsq >The transformation will be stored in .LSQ_RT_ b_to_a
 O > lsq_i
Lsq >Least squares match by Semi Automatic Alignment.
Lsq >There is an alignment called B_TO_A
Lsq >Given 2 molecules A,B the transformation rotates B onto A
Lsq >What is the name of molecule A [A      ]?
Lsq >Zone to look for alignment [all molecule A] :
Lsq >What is the name of molecule B [B      ]?
Lsq >Zone to look for alignment [all molecule B] :
Lsq >What atom [CA] ?
Lsq >Number of atoms in A/B to look for alignment   131   131
Lsq >Search for connected fragments.
Lsq >A fragment of   131    1    1 residues located.
Lsq >Loop =    1 ,r.m.s. fit =     0.614 with   131 atoms
Lsq >x(1) =      0.3866*x+    0.0907*y+    0.9178*z+   -16.9487
Lsq >x(2) =      0.1155*x+    0.9826*y+   -0.1458*z+    25.7526
Lsq >x(3) =     -0.9150*x+    0.1624*y+    0.3694*z+    66.9966
Lsq >Search for connected fragments.
Lsq >A fragment of   131    1    1 residues located.
Lsq >Loop =    2 ,r.m.s. fit =     0.614 with   131 atoms
Lsq >x(1) =      0.3866*x+    0.0907*y+    0.9178*z+   -16.9487
Lsq >x(2) =      0.1155*x+    0.9826*y+   -0.1458*z+    25.7526
Lsq >x(3) =     -0.9150*x+    0.1624*y+    0.3694*z+    66.9966
Lsq >The transformation can be stored in O.
Lsq >A blank is taken to mean do not store anything
Lsq >The transformation will be stored in .LSQ_RT_ b_to_a
Lsq >Here are the fragments used in the alignment
Lsq >    A1 SNKFLGTWKLVSSENFDEYMKALGVGLATRKLGNLAKPRVIISKKGDIIT
Lsq >    B1 SNKFLGTWKLVSSENFDEYMKALGVGLATRKLGNLAKPRVIISKKGDIIT
Lsq >    A51 IRTESPFKNTEISFKLGQEFEETTADNRKTKSTVTLARGSLNQVQKWNGN
Lsq >    B51 IRTESPFKNTEISFKLGQEFEETTADNRKTKSTVTLARGSLNQVQKWNGN
Lsq >    A101 ETTIKRKLVDGKMVVECKMKDVVCTRIYEKV    A131
Lsq >    B101 ETTIKRKLVDGKMVVECKMKDVVCTRIYEKV    B131
Lsq >File LSQ_PAINT.DAT will colour matched residues RED.
 O > !see it worked
 O > pair_at
Lsq >There is an matched pair called B_TO_A
Lsq >Object state ([ON],OFF) : on
 O > ! that shows some litle arrows betyween matched atoms
 O > stop
```

Appendix 2. An example of a non-trivial macro

The following macro was used on our work on cellulase CBH2 from T. reesei. This structure has two molecules in the asymmetric unit (we call them 'A' and 'B'). The macro expects us to identify an atom in molecule A, it will then display contours around this point from molecule A from and the equivalent point in molecule B. It also draws the atoms in molecule A that are 10 Å from the identified atom.

```
centre_id Wait_id
message 'Start contouring B'
copy 'save_centre' '.active_centre'
rot_trans_pb '.active_centre' 'a_to_b' '.active_centre'
map_file [alwyn.cbh2.ps300]b.map
map_obj mapb
map_par 12. 12. 12. 52. 32010 .5 .1 0
map_act
map_draw
rot_tran_obj b_to_a mapb
message 'Start contouring A'
copy '.active_centre' 'save_centre'
map_file [alwyn.cbh2.ps300]a.map
map_obj mapa
map_par 12. 12. 12. 52. 22010 .5 .1 0
map_act
map_draw
mol_al obj spha spher 10. end
message '  '
```

The first line waits until the user identifies an atom. The coordinates of an identified atom are stored in the data base at '.active_centre'. The message is sent to assure the user that something is happening. The coordinate is then saved somewhere else in the data base. The '.active_centre' is then transformed by a rotation/translation operator stored in 'a_to_b'. This operator could have been produced by the _Lsq menu or otherwise. The map data set to be used, and the object to be created are then defined. The 'map_par' command defines a cube of density 12x12x12 Å, to be contoured at a level of 52 units with a PS300 colour code equivalent to light blue (symbol definition is available if you don't like numbers). The front/back intensities are .5 and .1, and the '0' specifies the file type. The 'map_act' command uses the coordinate in '.active_centre' for the coordinate of the centre of the density. This is then drawn in object 'mapb'. This object is then transformed by 'b_to_a'.

The process is then repeated for the density around molecule A, after first restoring the initial '.active_centre'. Atoms from molecule 'A1' that are 10 Å from the centre, are then drawn into an object 'spha'.

This macro would be shown on the screen and could be activated by picking.

Inhibitor Design from Known Structures

RENEE L. DESJARLAIS, BRIAN SHOICHET,
GEORGE SEIBEL, and IRWIN D. KUNTZ, JR.

Most drugs exert their pharmacologic effect by binding to receptors. The receptor for a given drug may be an enzyme or other protein, DNA, or even the cell membrane. As knowledge of both drugs and disease states has grown, specific receptors have been identified for certain drugs, and target receptors have been identified for some disease states. Any effort to design drugs rationally depends on understanding the intermolecular interactions involved in the drug/drug receptor complex.

The design of a novel pharmaceutical agent is a long and complicated process. First, a lead compound with the desired activity is found. The discovery of a lead is typically made by random screening. More recently, the development of active compounds has been based on knowledge of receptor function (Cushman, *et al.*, 1977). Once a lead is found, it is modified and the derivatives are tested. The results of these tests are compiled into structure activity relations (Hansch, 1969) that can be used to predict modifications that will produce the derivative with optimal properties. In order for a compound to be a successful drug, the absorption, distribution, metabolic profile, and toxicity must be considered. Finally, the compound will be tested in animals and the clinic. There is much experience in the pharmaceutical industry with taking a lead compound through the development process to clinical trials, but the discovery of a novel lead is often primarily a matter of random screening and good luck. In theory, one could use the three-dimensional structure of a receptor to design molecules that would have high binding affinity for that receptor. These molecules could well be quite unusual and might then be used as novel leads in the drug discovery process. Structures of some medicinally interesting receptors are now available.

We have been developing a computer-assisted method for designing molecules complementary to a specific receptor. The design procedure begins by using the DOCK programs (DesJarlais, *et al.*, 1988, Kuntz, *et al.*, 1982) to find a set of molecules or molecular fragments that match the steric requirements of a receptor site. First, the shape of the receptor site is characterized. The molecular surface as described by Richards (1977) is the basis for this characterization. The program MS (Connolly, 1981, 1983a, 1983b) is used to generate a dot representation of this surface. Next, we build a negative image of the surface by finding a set of spheres that meet the following constraints: they must touch the surface at two points, must not overlap any receptor atoms, and must have their center along the surface normal at one of the surface points. The smallest spheres are the size of the spherical probe used in the generation of the molecular surface and the largest spheres are limited to 5 Å in radius. The number of spheres is also reduced by retaining only the largest sphere associated with a particular atom. Finally, the spheres are grouped into sets in which spheres overlap each other. Each of these sets of spheres characterizes a depression in the receptor surface. An average enzyme or small protein might have 3-5 such depressions that are of reasonable size to bind a small molecule, any of which may be used in the rest of the calculation. The depression with the largest number of spheres is typically the enzyme active site.

The next step in the procedure requires that we screen a database of small molecules for those whose shape best fits into the negative image described above. The database that we have used is a subset of the Cambridge Structural Database (Allen, *et al.*, 1979) consisting of approximately 10,000 molecules. These molecules have been selected to have a wide variety of shapes (Seibel, manuscript in preparation). Various orientations of each small molecule in the site are found by matching atom-atom distances from the small molecule to sphere-sphere distances from the receptor. Several hundred orientations are typically explored for each small molecule. The orientations are scored based on a simple function which approximates a soft van der Waals potential. The best orientation of each small molecule in the database is compared to the scores of other molecules. The top 50-200 molecules from the database are saved for further evaluation. For more details on the procedure, see DesJarlais, *et al.*, 1988.

We emphasize that the result of this search is a set of molecules that should have good van der Waals interaction with the receptor site, but we have up to now ignored other chemical properties of the receptor and the small molecule. Of course, realistic design requires examination of the electrostatic and hydrogen bonding properties of the receptor. The molecules selected from the database are considered templates. They require modification to assure chemical complementarity with the receptor site. As a first step in evaluating the chemical properties of the receptor, each atom of the small molecule in its highest scoring orientation is used to define a location at which the electrostatic potential from the protein atoms is evaluated. This potential is calculated using the partial charges in the AMBER united atom force field. (Weiner, *et al.*, 1984). Next, potential hydrogen bonds are identified. Oxygen and nitrogen atoms in the protein are located, and if there is a small molecule atom within hydrogen bonding distance of one of the protein oxygens or nitrogens, it is labeled as a potential hydrogen bond donor or acceptor. In our current program, the design process now becomes interactive. Properties of the receptor are highlighted as the user views the molecules in the receptor site using the molecular graphics program MIDAS (Ferrin and Langridge, 1980). The electrostatic potential at each small molecule is indicated by the color of the atom, and the potential hydrogen bonds are displayed as lines drawn from the oxygens and nitrogens in the protein to atoms in the small molecule that are within hydrogen bonding distance (DesJarlais, et al,. 1989, and DesJarlais, 1987). It is useful to display the potential hydrogen bonds as lines, because this emphasizes the angular component of the hydrogen bonding interaction. In collaboration with organic and medicinal chemists, the molecular templates are examined in detail and specific modifications are proposed to achieve both good chemical matches and synthetic feasibility.

The energy of interaction between the modified molecules and the receptor can be explored with molecular mechanics programs such as AMBER (Weiner and Kollman, 1981).

The design procedure described above was applied to the protein penicillopepsin, an aspartyl protease from *Penicillium janthinellum*. This protein is related both functionally and structurally to the protease produced by the human immunodeficiency virus. The coordinates for penicillopepsin were obtained from the Protein Data Bank (Bernstein, *et al.*, 1977) entry 2APP. The crystal structure was determined by James and Sielecki (1983) at 1.8 Å resolution. The DOCK package of programs was used to obtain molecules that have a good geometric fit to the enzyme active site. One of the molecules found in the shape search will be discussed in detail to explain the evaluation of the electrostatic and hydrogen bonding interactions with the receptor.

Figure 1a shows a molecule that was found in the shape search to be complementary in shape to the active site of penicillopepsin. This molecule, 2-(2-quinolyl) cyclohexane phenylhydrazone, (Bocelli, *et al.*, 1984) was ranked fifteenth in the shape fit. It is more interesting than the higher scoring molecules because its framework is fairly simple from a synthetic standpoint. This molecule fits tightly into the site with its quinolyl end in a deep channel that extends into the protein (see Figure 1a). The cyclohexane ring gives the molecule an over all "L" shape, and the phenyl portion lies along the active site groove near the active site aspartyls. Figure 1b shows the electrostatic potential from the protein at each of the small molecule atom centers. The potential ranges from -100 kcal/mole per unit electron charge (red) to 0 kcal/mole per unit electron charge (green) with intermediate values shaded intermediate colors (e.g. yellow at about -50 kcal/mole per unit electron charge). Receptor polar atoms that might require hydrogen bonds when the molecule is docked with the proposed geometry are shown in Figure 1c, with lines drawn from the receptor atoms to nearby ligand atoms.

The design of hydrogen bonding interactions is not a simple problem. Because hydrogen bonds are directional, the position of the small molecule atom with respect to the protein atom is of more concern than for electrostatic interactions. We are working on a procedure for assessing how well each small molecule might be able to satisfy the hydrogen bond requirements of the protein atoms that it is near. We expect that this will allow identification of the most promising small molecules early in the screening and reduce the amount of time spent examining molecules that cannot make the necessary interactions with the receptor.

Beyond the problem of whether a particular interaction will be favorable is the question of how many potential interactions must be made for a ligand to bind tightly. Experimental data indicate that a single hydrogen bond can be worth 0.5-1.5 kcal/mole in binding energy. (Fersht, et al., 1985) It would be desirable to satisfy as many of these interactions as the protein would seem to require, but it may not be necessary to make them all. X-ray crystallographic data on the binding of antiviral compounds to human rhinovirus (Smith, et al., 1986) shows that although the compound is virtually surrounded by protein, only one hydrogen bond is made leaving some protein atoms and some atoms from the antiviral compound without hydrogen

---

See color insert:

Figure 1: A small molecule found in the shape search is shown in the active site of Penicillopepsin. The enzyme and its molecular surface are shown in blue with the active site aspartates in red. (a) The van der Waals surface of the small molecule (orange) is shown to illustrate the good fit. (b) The same small molecule with its atoms colored by the value of the electrostatic potential from the protein. Red indicates a potential of -100 kcal/mole per unit electron charge and green indicates a potential of 0 kcal/mole per unit electron charge. The color of intermediate values of the potential are scaled accordingly. (c) The same small molecule with potential hydrogen bonds to the protein shown as lines between the protein atoms and small molecule atoms. Pink lines indicate that the protein atom is a carbonyl oxygen. Green lines indicate that the protein atom is a hydroxyl oxygen. Purple lines indicate that the protein atom is a nitrogen. Reprinted with permission from DesJarlais, R. L.; Seibel, G. L.; Kuntz, I. D. "A Second Generation Computer-Assisted Inhibitor Design Method" in Bioactive Mechanism: Proof, SAR, and Prediction; Magee, P.,Block, J., and Henry, D., Eds.; ACS Symposium Series; American Chemical Society: Washington, DC, 1989, in press. Copyright 1989 American Chemical Society

bond partners. Clearly, one would like to make as many favorable interactions as possible for the best binding affinity, but one must also be aware of the potential for designing in interactions that cannot all be made simultaneously.

The DOCK package of programs is able to find small molecules with shape complementarity to a given receptor site. The use of interactive molecular graphics to display potential hydrogen bonds and the electrostatic potential has proven to be a useful guide in determining what modifications might be made to increase chemical complementarity with the receptor site. Considerable amounts of time are required to do this, however. For the molecules that look promising, it may take several hours at an interactive graphics device to design a target molecule. Moreover, it often takes significant effort to decide that a molecule cannot complement the chemistry of the active site adequately and is not a good candidate.

The ultimate test of any design method will come in the laboratory. Synthesis of penicillopepsin inhibitors as well as inhibitors of other enzyme systems is on going and testing will begin when molecules are in hand. These studies will involve crystallographic examination as well as conventional binding studies. We expect that the results of these experiments will allow us to improve our design method and better understand intermolecular interactions in general.

A second area of interest to us is the prediction of the favorable co-positioning of two interacting biological molecules. This remains a difficult problem in biochemistry. Work has normally proceeded using active sites whose relationship to the putative ligands has been largely established by crystallography (Goodford, 1985; Cody, 1986, Bash et al., 1987). The more general issue of predicting how a given ligand will bind to a macromolecule when a structural model for such binding is unavailable has received less attention, owing to the computational problems presented by the complex energy surfaces of biological macromolecules.

We have used the shape-fitting algorithms to overcome some of the computational difficulties of modeling protein-ligand interactions when the structure of the model complex is not known. Previous work has shown that the docking method is able to regenerate the crystallographically determined orientations of heme in hemoglobin (Kuntz et al., 1982), methotrexate in dihydrofolate reductase (DesJarlais et al., 1986), NADH in lactate dehydrogenase and the configurations of several other protein-ligand systems (Shoichet and Kuntz, unpublished results). Here, we present an approach for addressing the question of modeling protein-ligand interactions in the case of predicting the ternary complex of *L. casei* Thymidylate Synthase (TS) with its two natural substrates, deoxyuridine monophosphate (dUMP) and methylenetetrahydrofolate (CH_2-H_4folate). This work used the 3.0 Å crystal structure of unliganded *L. casei* TS (Hardy et al., 1987) and of the uncomplexed structures of the substrates.

We began with the uncomplexed structure of each molecule, TS, CH_2-H_4folate and dUMP. We used four of the most energeticaly stable conformers of dUMP and four conformers of the folate as starting geometries for the ligands. The various dUMP conformers were docked into TS using the method of Kuntz (above), generating starting configurations in the enzyme. These starting structures were energy-minimized using the molecular mechanics program AMBER (Weiner and Kollman,1981). Using this process we generated non-covalent binary complexes between TS and dUMP and then proceeded to covalent complexes by introducing a format carbon-sulfur bond and reminimizing. Further docking and minimization involving the binary complexes and the folate conformers generated a set of TS-dUMP-CH_2-H_4folate partially covalent ternary complexes involving a covalent bond between TS and dUMP and non-covalent interactions between the enzyme and the CH_2-H_4folate.

In this manner we have generated four possible ternary complex structures. All four structures have a covalent bond between the SG of Cys 198 of TS and the C6 of dUMP and the phosphate of dUMP near Arg 218 and Arg 179'. The dihydro-pyrimidine is in a half-chair conformation, with the SG and the CH_2-H_4folate in a trans-diaxial relationship. Two possible stereochemistries are predicted for the dUMP in the fully covalent ternary complex: either (5S,6R) or (5R,6S). In the former, the CH_2-H_4folate lies parallel to the major axis of the dUMP in either of two possible orientations, while in the latter the CH_2-H_4folate lies orthogonal to the major axis of the dUMP, again in either of two possible orientations. The modeling suggests specific roles for certain TS residues in the binding of the substrates. Our predictions are listed Table I along with the results of the experimental tests that have performed as of this writing.

TABLE I
PREDICTIONS AND TESTS

PREDICTION	TEST	RESULTS
dUMP in ternary complex is one of two stereoisomers: 1) (5S,6R) or 2) (5R,6S).	Solution of ternary complex structure.	
CH_2-H_4folate configurations depend on dUMP stereochemistry: in 1) the folate is oriented perpendicular to the dUMP, in 2) the folate is oriented parallel to the dUMP.	Solution of ternary complex structure.	
His 199 has no *direct* role in catalysis.	Site-directed mutagenesis of His to a non-basic amino acid.	His mutated to Val: enzyme retains 25% of specific activity.[a]
Tyr 261 hydrogen bonds to 3'-OH of dUMP.	Mutate Tyr to Phe: K_d should increase.	Most Y261 mutants are nonviable[b]
His 259 hydrogen bonds to either: 1) dUMP phosphate or 2) Ser 219	Mutate His to Val or Phe: K_d goes up, probably more in 1) than 2).	
Ser 219 hydrogen bonds to either: 1) dUMP phosphate or 2) 3'-OH of dUMP	Mutate Ser to Ala: K_d goes up in either case.	
Trp 82 interacts with CH_2-H_4folate	Mutate Trp to Phe: Study affect on UV absorbance of folate binding.	
Asp 221 interacts with Tyr 261 in hydrogen bond network.	Mutate Asp to Val or Ser: K_d of dUMP should increase: temperature/binding studies should show affect on $\Delta\Delta S$ of binding.	
Asp 221 may also interact with CH_2-H_4folate. This interaction only found in two of the four putative structures.	Mutate Asp to Val or Ser K_d of folate should increase.	
Tyr 146 hydrogen bonds with O4 of dUMP found in (6R) binary complexes, not in ternary complexes	Mutate Tyr to Phe Effect on K_d, K_{cat} of dUMP	

a. Frasca et al., 1988.
b. Dr. Shane Climie, personal communication.

The most complete test of these predictions will be the solution of a TS covalent ternary complex crystal structure.

208

In conclusion, the docking procedures described here can be used to propose novel chemical structures as putative inhibitors of enzymes of known tertiary structure and can also be used to model enzyme-substrate-cofactor complexes. Experimental tests of these ideas are in progress.

ACKNOWLEDGMENTS. We wish to thank our many collaborators: Dale Bodian of the University of California, San Francisco, Robert Sheridan of Lederle Laboratories, and J. Scott Dixon of Smith Kline & French Laboratories, as well as Paul Ortiz de Montellano and Marlyse Gander. Special thanks to Peter Kollman and his research group for help with AMBER and to Robert Langridge and the Computer Graphics Laboratory for help with MIDAS. Support for this project comes from GM-39552 (G. L. Kenyon, Principal Investigator) and GM-31497 (I.D.K.). Additional support from the Defense Advanced Research Projects Agency under contract N00014-86-K0757 administered by the Office of Naval Research has been most helpful. The University of California, San Francisco Computer Graphics Laboratory is a National Institutes of Health Research Resource (RR-01081)

REFERENCES

Allen, F.H.; Bellard, S.; Brice, M. D.; Cartwright, B. A.; Doubleday, A.; Higgs, H.; Hummelink, T.; Hummelink-Peters, B. G.; Kennard, O.; Motherwell, W. D. S.; Rodgers, J. R.; Watson, D. G. "The Cambridge Crystallographic Data Centre: Computer-Based Search, Retrieval, Analysis, and Display of Information", Acta Crystallogr., Sect.B 1979, B35, 2331.

Bash, P.A, Singh, U.C., Brown, F.K., Langridge, R. and Kollman, P.A., "Calculation of the Relative Energy of a Protein-Inhibitor Complex", Science 1987, 235, 574-576.

Bernstein, F. C.; Koetzle, T. F.; Williams, G. J. B.; Meyer, E. F. Jr.; Brice, M. D.; Rodgers, J. R.; Kennard, O.; Shimanouchi, T.; Tasumi, M. "The Protein Data Bank: A Computer-based Archival File for Macromolecular Structure", J. Mol. Biol. 1977, 112, 535.

Bocelli, G.; Tosi, G; Cardellini, L. "2-(2-Quinolyl)cyclohexanone Phenylhydrazone, $C_{21}H_{21}N_3$", Acta Cryst. 1984, C40, 1952.

Cody,V., "Computer Graphic Modelling in Drug Design" J. Mol. Graph. 1986, 4(1), 69-73.

Connolly, M. L. "Protein Surfaces and Interiors", Ph.D. Thesis, University of California, Berkeley,1981.

Connolly, M. L. "Analytical Molecular Surface Calculation", J. Appl. Crystallogr. 1983a, 16, 548.

Connolly, M. L. "Solvent-Accessible Surfaces of Proteins and Nucleic Acids", Science 1983b, 221, 709.

Cushman, D. W.; Cheung, H. S.; Sabo, E. F.; Ondetti, M. A. "Design of Potent Competitive Inhibitors of Angiotensin-Converting Enzyme. Carboxyalkanoyl and Mercaptoalkanoyl Amino Acids", Biochemistry 1977, 16, 5484.

DesJarlais, R. L.; Seibel, G. L.; Kuntz, I. D. "A Second Generation Computer-Assisted Inhibitor Design Method" in Bioactive Mechanism: Proof, SAR, and Prediction; Magee, P.,Block, J., and Henry, D., Eds.; ACS Symposium Series; American Chemical Society: Washington, DC, 1989, in press.

DesJarlais, R. L.; Sheridan, R. P.; Seibel, George L.; Dixon, J. S.; Kuntz, I. D.; Venkataraghavan, R. "Using Shape Complementarity as an Initial Screen in Designing Ligands for a Receptor Binding Site of Known Three-Dimensional Structure", J. Med. Chem. 1988, 31, 722.

DesJarlais, R. L. "Molecular Shape Complementarity:A Method For Finding New Lead Compounds", Ph. D. Thesis, University of California, San Francisco, 1987.

DesJarlais, R.L., Sheridan, R.P., Dixon, S.J., Kuntz, I.D. & Venkataraghavan, R. "Docking Flexible Ligands to Macromoleclar Receptors by Molecular Shape", J. Med. Chem. 1986, 29, 2149.

Ferrin, T. E. and Langridge, R. "Interactive Computer Graphics with the UNIX Time-Sharing System", Computer Graphics 1980, 13, 320.

Fersht, A. R.; Shi, J.-P.; Knill-Jones, J.; Lowe, D. M.; Wilkinson, A. J.; Blow, D. M.; Brick, P.; Carter, P.; Waye, M. M. Y.; Winter, G. "Hydrogen Bonding and Biological Specificity Analysed by Protein Engineering", Nature 1985, 314, 235.

Frasca, V., LapPat-Polasko, L., Maley, G.F. & Maley, F. "Site-Directed Mutagenesis of the T4-Phage Thymidylate Synthase Gene" Advances in Gene Technology: Protein Engineering and Production, Proceedings of the Miami Winter Symposium, 1988, 149.

Goodford, P.J "Drug Design by the Method of Receptor Fit" J. Med. Chem 1984, 27(5), 551.

Hansch, C. "A Quantitative Approach to Biochemical Structure-Activity Relationships", Acc. Chem. Res. 1969, 2, 323.

Hardy, L.W., Finer-Moore, J.S., Montfort, W.R., Jones, M.O., Santi, D.V. & Stroud, R.M., "Atomic Structure of Thymidylate Synthase: Target for Rational Drug Design" Science 1987, 235, 448-455.

James, M. N. G.; and Sielecki, A. R. "Structure and Refinement of Penicillopepsin at 1.8 Å Resolution", J. Mol. Biol. 1983, 163, 299.

Kuntz, I. D.; Blaney J. M.; Oatley S. J.; Langridge R.; Ferrin, T. E. "A Geometric Approach to Macromolecule-Ligand Interactions", J. Mol. Biol. 1982, 161, 269.

Richards, F. M. "Areas, Volumes, Packing, and Protein Structure", Annu. Rev. Biophys. Bioeng. 1977, 6, 151.

Smith, T. J.; Kremer, M. J.; Luo, M.; Vriend, G.; Arnold, E.; Kamer, G.; Rossmann, M. G.; McKinlay, M. A.; Diana, G. D.; Otto, M. J. "The Site of Attachment in Human Rhinovirus 14 for Antiviral Agents That Inhibit Uncoating", Science 1986, 233, 1286.

Weiner, P. K.; and Kollman, P. K. "AMBER: Assisted Model Building with Energy Refinement. A General Program for Modeling Molecules and Their Interactions" J. Comp. Chem. 1981, 2, 287.

Weiner, S. J.; Kollman, P. A.; Case, D. A.; Singh, U. C.; Ghio, C.; Alagona, G.; Profeta, S.; Weiner, P. "A New Force Field for Molecular Mechanical Simulation of Nucleic Acids and Proteins", J. Am. Chem. Soc. 1984, 106, 765.

The Cambridge Structural Database in Molecular Modeling: Systematic Conformational Analysis from Crystallographic Data

FRANK H. ALLEN and MICHAEL J. DOYLE

Introduction

The investigation of low-energy conformations of molecules or substructures is fundamental to the process of molecular modelling. A number of computational procedures exist for this purpose, but suffer from some operational or scientific limitations. These primarily concern the size of molecules that can be processed, availability of relevant and reliable force-field parameters, or on the need to postulate a number of starting geometries so as to scan the whole of conformational space. It is reasonable to assume, however, that conformations observed in crystal structures are close to one or more minima on the potential energy hypersurface. The Cambridge Structural Database (CSD: Allen et al., 1984) now contains 3D crystallographic data for some 73000 organocarbon compounds. It represents an ever-growing compendium of conformational information over a very broad chemical spectrum: a common chemical substructure may well occur in several hundred of these crystal structures. It is therefore necessary to develop rapid automatic techniques by which such large datasets may be sorted into conformational subgroups. These subgroups may then be ranked in order of their population; if two or more well-populated subgroups exist, then a representative or averaged conformation from each may be used as an (energetically accessible) alternative in model building.

This paper reviews current progress in the development of algorithms for 3D pattern recognition. These techniques are used to identify conformational subgroups in large datasets of crystallographic results relating to a specified substructure.The results can be presented graphically and numerically for rapid assimilation and use in a molecular modelling environment. These algorithms are being developed within current Versions 3 and 4 of the CSD System, which are briefly summarised below.

The Cambridge Structural Database System

The CSD System comprises the database itself together with associated software for search, retrieval, analysis and display of stored information. The database covers all X-ray and neutron diffraction studies of organics, organometallics and metal complexes. In June 1989 there were 73983 entries, a figure which increases by some 7000 new entries per year. Three categories of information are held for each entry: (a) bibliographic and chemical text and some individual numerical items; (b) a full 2D chemical connection table, recently upgraded (Version 4) with x, y-atomic coordinates for the preparation of 2D chemical structural diagrams; (c) 3D crystallographic coordinates, symmetry etc. Total database size is ca. 160Mb, with some 65% of information being in category (c).

The CSD consists of a single unified file (ASER: Figure 1), which is specially structured for rapid searching, via the addition of some 700 bit screens (Allen et al., 1988). Access to the 3D data is currently via a two stage process (Figure 1). Information in categories (a), (b) above is searched using program QUEST. A wide variety of text, numerical and chemical fields may be interrogated individually or in (Boolean) combinations. The interrogation of the 2D connection tables to locate substructural fragments allows for very flexible definition in chemical terms. Generic queries are simple to code and 2D similarity searching (Willett et al., 1986) is also supported. User input to QUEST is alphanumeric (Version 3), or via a menu-driven graphics interface (Version 4) with output of chemical structural diagrams.

The 3D crystallographic data for hits generated by QUEST are retrieved as a subfile (FDAT: Figure 1). This file is then processed by PLUTO to generate 3D diagrams and illustrations, and by GSTAT for numerical and statistical analyses. GSTAT contains extensive 3D substructure search facilities for both intra- and inter-molecular fragments. Most importantly, the program will generate systematic tabulations of user-specified geometrical parameters for these substructures. Simple statistics, eg means and standard deviations, are provided for these quantities, which may also be displayed via simple histograms and scattergrams. Mean dimensions for substructures are readily obtained, and

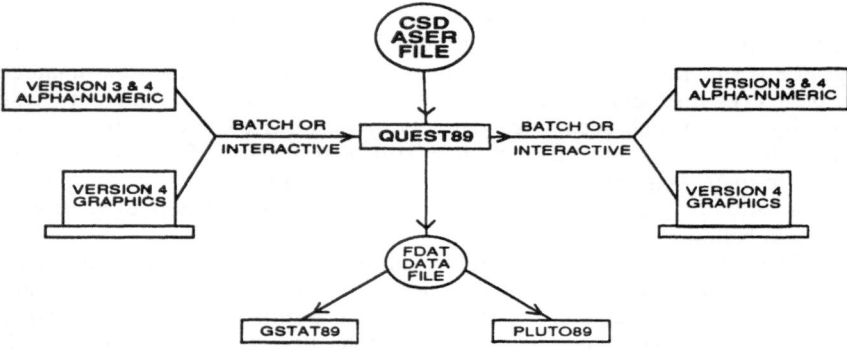

Figure 1. Flowchart of the Cambridge Structural Database System.

are also of considerable use in model building. New compilations of 'standard' bond lengths for organics (Allen et al., 1987) and metal complexes (Orpen et al.,1989) have been prepared by this route. The GSTAT tabulation and simple graphics can also be used to study the dimensions and directionality of H-bonded interactions (Taylor & Kennard 1984) and other non-bonded contacts. This is vital information for modelling purposes and the topic is covered in greater depth in another Chapter in this Volume.

The fragment geometry tables may also be analysed by more sophisticated statistical methods within GSTAT. These include correlation and regression techniques and principal component analysis. The cluster analysis algorithms described below represent a further extension of the statistical functionality in GSTAT.

Conformational Analysis via CSD: An Overview

Experimentally observed conformations for a given substructure can be expressed in terms of Nt torsion angles for the Nf occurences of the fragment. A raw data matrix of this type can readily be derived from CSD using the systematic tabulation facilities of the GSTAT program. In some cases the fragment will have only one (Figure 2a) or two (Figure 2b) degrees of torsional freedom . Here the univariate or bivariate distributions of Nt over the Nf fragments are simply displayed as histograms or scattergrams. Figure 2a clearly shows that cyclopropyl-carbonyls prefer the cis ($\tau \approx 0°$) or trans ($\tau \approx 180°$) bisected conformations. Figure 2b shows that acyclic primary esters prefer a synplanar-antiplanar arrangement with $\tau_1/\tau_2/\approx 0,180°$ For most practical applications, particularly those involving ring systems, Nt >> 2 and the raw data matrix is now multivariate. Two main techniques have been applied to the analysis of multivariate distributions of torsion angles.

Principal component analysis (Murray-Rust et al., 1978a,b; 1985) can be used to construct the Mt mutually orthogonal principal components (linear combinations of the original Nt torsion angles) which account for a large proportion of the variance of the multivariate dataset. In many cases Mt<<Nt and the dimensionality of the problem is reduced. Pairs of the Mt principal component scores for the Nf fragments may thus be plotted as 2D scattergrams, conformational subgroups are identified by visual inspection. Application of this technique to the β–1'- aminoribofuranosyl fragment (Murray-Rust et al., 1978b, Taylor 1986a) showed a clear preference for the 2'-endo and 3'-endo conformations. Occasionally the principal component axes may be interpreted in chemical terms, especially for cyclic systems. We will return to this topic in a later section of this Chapter.

A variety of agglomerative clustering techniques have also been applied to multivariate torsional datasets (Murray-Rust et al., 1986; Norskov-Lauritsen et al., 1985; Taylor 1986a). The first step in these techniques is to calculate the conformational dissimilarity of each pair of observations in the dataset. This information is used to break down the complete distribution into clusters, each cluster containing fragments of similar conformation. The results may be presented as numerical tabulations of torsion angles for each discrete cluster, from which an average conformation can be derived. Taylor (1986a) also describes a number of other possible techniques for multivariate analysis.

214

```
              5    10   15   20   25   30   35   40   45   50      Nf
         ....I....I....I....I....I....I....I....I....I.....
  -10.0 -                                                    -
         .*******************************                  . ( 33)
         .********************************************************. ( 49)
         .**********************                            . ( 23)
         .************                                      . ( 12)
   40.0 -********                                           - (  8)
         .****************                                  . ( 16)
         .*******                                           . (  7)
         .****                                              . (  5)
         .******                                            . (  6)
   90.0 -****                                               - (  5)
         .
         .*****                                             . (  5)
         .**                                                . (  2)
         .****                                              . (  4)
  140.0 -****                                               - (  4)
         .******************                                . ( 18)
         .**************************                        . ( 26)
         .****************                                  . ( 17)
         .****************                                  . ( 16)
  190.0 -                                                    -
         ....I....I....I....I....I....I....I....I....I.....
  TAU    5    10   15   20   25   30   35   40   45   50      Nf
```

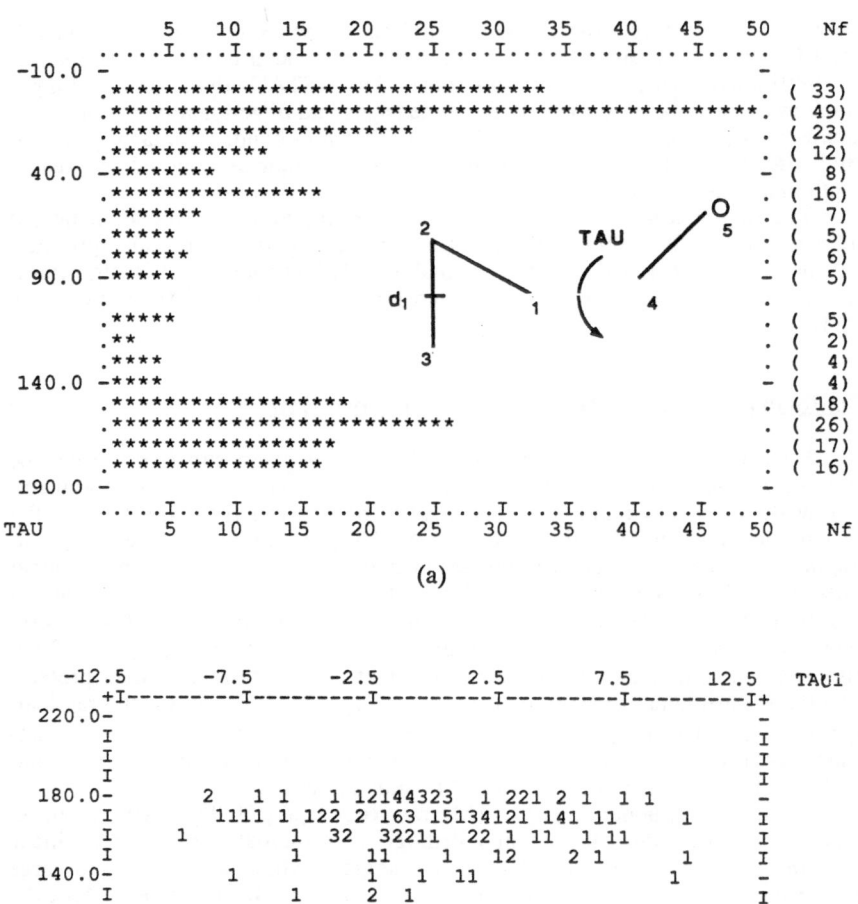

(a)

```
   -12.5         -7.5         -2.5          2.5          7.5         12.5   TAU1
     +I---------I----------I----------I----------I----------I----------I+
 220.0-                                                                    -
     I                                                                     I
     I                                                                     I
     I                                                                     I
 180.0-      2    1 1    1 12144323  1 221 2 1   1 1                        -
     I          1111 1  122 2  163 15134121 141 11       1                 I
     I      1          1  32   32211  22 1 11   1 11                        I
     I                 1        11   1   12    2 1            1             I
 140.0-          1          1    1   11                      1             -
     I                    1      2 1                                        I
     I                  1 1      1 1 1                        1             I
     I                  1     1 1  11    2  1    1                          I
 100.0-                  11        1 1    2   1                    1        -
     I 1                1 1 1     121  12221              1                 I
     I          1  1 11 1     12   11     1 1 11 11             1           I
     I                1            1                                        I
  60.0-                                                                    -
     I                                                                     I
     I                                                                     I
     I                                                                     I
  20.0-                                                                    -
     +I---------I----------I----------I----------I----------I----------I+
 ABS(TAU2)                                                            Nf=191
```

(b)

Figure 2. (a) Distribution of the absolute value of TAU - the torsion angle
O(5)-C(4)-C(1)-d1 where d1 is the mid point of C(2)-C(3). (b)
Scattergram of TAU1 (O=C-O-CH$_2$) versus the absolute value of TAU2
(C-O-CH$_2$-C).

The introduction of these techniques represents an enormous step towards automatic conformational analysis from large datasets of crystallographic results. The initial implementaions of these algorithms do, however, suffer from some deficiencies in their application in this area. Both techniques have been shown to work well when applied to fragments which are asymmetric, however many small but interesting fragments exhibit 2D symmetry in their chemical structure. This topological symmetry can cause severe dificulties in interpreting results in terms of the underlying conformational minima These difficulties are exemplified by a principal component analysis of phosphate groups (Murray-Rust, 1982) and a cluster analysis of bis (triphenylphosphine)-metal complexes (Norskov-Lauritsen et al., 1985). Secondly, implementation of both techniques falls just short of the desired aims for molecular modelling purposes, as identified in the Introduction: the ranking of conformations in order of importance, and the provision of atomic coordinates representative of each major conformer.

We now show how these remaining difficulties can be overcome, and describe how the two techniques above are being combined into a potentially powerful package for conformational studies via CSD. For example purposes we use a trial dataset of Nf=222 six-membered carbocycles with conformations defined by the Nt=6 intra-annular torsion angles $\tau_1-\tau_6$. The dataset was selected from CSD to include conformational variety in the form of (a) cyclohexanes (chairs), (b) norbornanes (boats), (c) cyclohex-1-enes (half-chairs, sofas, etc) and (d) phenyl rings (planar). The topological symmetry of the basic 2D substructure is D6h.

Effects of Fragment Symmetry

The effects of fragment symmetry for the sample dataset can be seen immediately in sections of the GSTAT tabulation of torsion angles shown in Table 1. Misalignment of the torsional sequences, for (a) boats and (b) chairs, is obvious. This arises from the atom-by-atom bond-by-bond mapping of the D6h-symmetric 2D fragment onto each 'target' molecule in CSD. If the fragment is numbered 1-6 cyclically, then there are 6 possible ways of mapping atom 1 onto each target. With atom 1 mapped, there are two possible ways of mapping atom 2, giving rise to 12 alternative 2D mappings. Furthermore, each 3D target has an enantiomorph of equal interest. For a given ring, the mapping will be chosen randomly, hence any one of a possible 24 permutations of the torsional sequence may occur in the GSTAT table.

A single-linkage cluster analysis routine (Taylor, 1986b) described in more detail below, was applied to the raw data matrix of torsion angles. Mean torsion angles and populations for the boat and chair clusters identified by the algorithm are in Table 2. The boat conformations (Table 2a) are spread over 5 of the 6 possible torsional permutations of Table 1a, the sixth permutation (fragment 131) is a representative of a cluster with a population of only 2.

The random nature of the atom mapping procedure is reflected by Np values in the range 2-15 covering 47 true boats in the sample. Similar remarks apply to the chairs (Table 2b) where 57 fragments are divided unequally between the two enantiomorphs.

Table 1. Representative torsional sequences generated by GSTAT for six-membered carbocycles.

Fragment	$\tau 1$	$\tau 2$	$\tau 3$	$\tau 4$	$\tau 5$	$\tau 6$
(a) Boats						
63	1.7	70.7	-73.6	2.2	70.2	-72.0
69	-1.2	-71.7	71.3	1.7	-70.8	70.4
114	-78.1	72.9	0.2	-69.8	64.4	5.2
121	70.1	-1.0	-72.5	72.8	-4.7	-65.8
131	-70.5	-0.6	71.1	-68.7	-2.5	72.7
134	68.2	-73.2	1.8	70.9	-75.6	4.7
(b) Chairs						
1	-60.5	59.7	-61.1	63.4	-60.4	58.9
9	56.1	-56.0	56.4	-53.9	54.7	-57.2

A principal component analysis of the raw torsional data yielded 3 principal components accounting for 99.9% of the total variance. The two scattergrams of Figure 3 provide a graphical illustration of the clustering results. The central peak in (a) is due to phenyl rings of approximate D6h symmetry, it is flanked at PC2=0 by two peaks of unequal height generated by the two chair enantiomorphs. The line of population density at PC1=0 is generated by the 6 permutational variants of the boat form rings. The view along PC1 in (b) shows that this line of density is resolved into a circle of peaks around the origin in the PC2, PC3 plane. The two large and four smaller maxima correspond to the boat conformations identified by the clustering procedure.

It is obvious, then, that fragment symmetry will always cause problems in the interpretation and use of results from clustering algorithms or principal component plots. Even if the fragment is topologically asymmetric in 2D, the 3D enantiomorphic pairs will always occur and must be considered together in

Table 2. Mean torsion angles (e. s. d.'s in parentheses) for boat and chair clusters identified by the single-linkage algorithm without symmetry modification. Nc is a cluster number, Np is the population of the cluster.

Nc	Np	$\tau 1$	$\tau 2$	$\tau 3$	$\tau 4$	$\tau 5$	$\tau 6$
(a) Boats							
1	15	4.8(17)	-75.5(11)	71.4(8)	1.1(11)	-70.8(10)	67.2(16)
2	15	0.1(9)	65.9(17)	-65.9(25)	-0.4(15)	66.7(17)	-66.6(22)
3	7	63.6(33)	0.5(17)	-63.5(31)	62.2(35)	0.5(18)	-63.4(19)
4	4	-70.6(26)	72.2(4)	-1.3(5)	-69.7(3)	69.6(24)	-0.4(21)
5	4	67.9(4)	-71.4(10)	0.8(6)	70.7(16)	-69.3(35)	0.7(21)
(b) Chairs							
6	38	54.3(9)	-53.7(8)	54.1(9)	-55.4(12)	55.4(9)	-54.7(7)
7	19	-53.9(15)	53.7(10)	-53.4(15)	-53.6(14)	-53.6(17)	53.1(20)

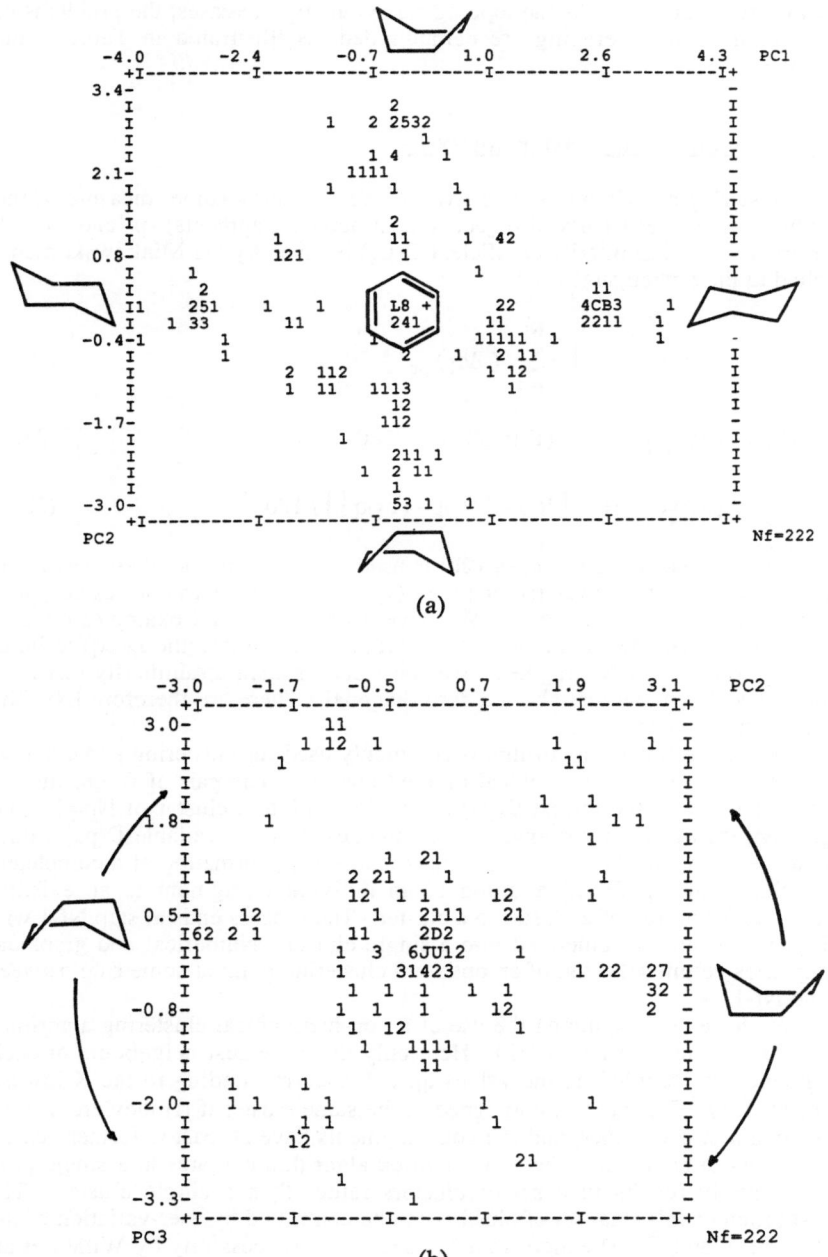

Figure 3. Principal component plots derived from the raw torsion angle table for six-membered carbocycles

any averaging process. As the topological symmetry increases, the problems of interpretation and averaging are compounded, as illustrated in Table 2 and Figure 3.

Symmetry-modified cluster analysis

Clustering algorithms (see eg Everitt, 1980) employ some measure of the dissimilarity between pairs of objects (substructural fragments) p and q. A conformational dissimilarity coefficient D(pq) is given by the Minkowski metric applied to the torsion angles :

$$D(pq) \quad = \quad \left[\ \sum_{i=1}^{Nt} \ (\Delta\tau_i)^n_{pq} \ \right]^{1/n} \tag{1}$$

$$\text{where} \quad (\Delta\tau_i)_{pq} \quad = \quad | \ (\tau_i)p - (\tau_i)q \ | \ /180 \tag{2a}$$

$$\text{or} \quad (\Delta\tau_i)_{pq} \quad = \quad \left[360 - | (\tau_i)p - (\tau_i)q | \right] / 180 \tag{2b}$$

The minimum value from (2a) or (2b) is used in (1) due to the phase restriction $-180 < \tau < 180$. The power factor (n) in (1) is an integer variable; usually n=1 (city block) or n=2 (Euclidean). We have used n=1, for all examples cited in this Chapter. The denominator (180) in (2a, 2b) normalizes the D(pq) to lie in the range 0 to 1. For Nf fragments we construct a square dissimilarity matrix of order Nf and symmetrical about a zero diagonal. There are therefore [Nf (Nf-1)]/2 unique D(pq) values.

The single-linkage algorithm is commonly used for clustering and employs the D(pq) values in a hierarchical ordered manner. The pair of fragments p,q which give rise to the lowest D(pq) are coalesced into a cluster of Np=2. The algorithm then continues stepwise, using the next lowest available D(pq) value. At each step one of the following actions is taken: (i) formation of a completely new cluster of Np=2; (ii) addition of an individual fragment to an existing cluster; (iii) addition of a cluster to a cluster. The process ends at step Nf-1 with all Nf fragments agglomerated into a single cluster. Numerical and graphical summaries permit selection of an optimum clustering point at some step between 1 and Nf-1.

We have also examined the use of a non-hierarchical clustering algorithm due to Jarvis and Patrick (1973). Here only the K nearest neighbours of each fragment are recorded, ie the values q_i, i=1,K corresponding to the K lowest D(pq) values. Fragments are assigned to the same cluster if (i) they are nearest neighbours of each other, and (ii) both fragments have at least C further nearest neighbours in common. The Jarvis-Patrick algorithm operates in a single pass and normally results in a set of clusters rather than a single cluster. The constitution of this final set of clusters can be controlled by user-variation of the values of K and C. The algorithm has been used successfully by Willett et al. (1986b) for clustering (classification) of large collections of 2D chemical structres on the basis of similarities in their connection tables. It has proved to be both computationally and operationally effective here in the clustering of 3D structures using torsional descriptors.

In order to circumvent the problems highlighted in the previous section, the possible symmetry-equivalent permutations of the original torsional sequences are specified to the program. The possible permutations are dictated by the full 2D topological symmetry of the fragment. A separate flag requests that conformational enantiomorphs (inversions of torsional sign sequences) are also to be considered. For the sample of six-membered carbocycles there are 12 permutations, including the identity, and the inversion flag is set on. The key to the symmetry-modification of clustering algorithms lies in the calculation of dissimilarities via equations 1, 2a, 2b. With the torsional sequence for fragment p held static, the sequence for q is allowed to adopt all possible permutations/inversions demanded by symmetry. The D(pq) value recorded in the matrix (single-linkage) or used to select nearest neighbours (Jarvis-Patrick) is the lowest value obtained over all symmetry variants of fragment q. The permutation of q which gives rise to min[D(pq)] is also stored as ±s, where s is in the range 1 to 12 for the sample dataset and + or - indicates use of the existing or inverted torsional sign sequence. By this means, the misalignments of torsional sequences exhibited in Table 1 are removed. These symmetry relationships ±s must, of course, be taken into account in later stage of each algorithm. This is particularly important for the agglomeration stages (i)-(iii) of the single-linkage method.

Table 3. Mean torsion angles (e. s. d.'s in parentheses) for major clusters identified by the symmetry-modified single-linkage algorithm. Nc is a cluster number, Np is the population of the cluster.

Nc	Np	$\tau 1$	$\tau 2$	$\tau 3$	$\tau 4$	$\tau 5$	$\tau 6$
(a) Phenyl							
1	35	1.1(1)	0.6(2)	-1.0(2)	-0.4(1)	2.1(2)	-2.4(3)
(b) Boats							
2	34	2.9(5)	68.3(5)	-69.5(6)	-0.3(4)	70.9(6)	-73.2(7)
3	11	1.0(4)	-57.5(6)	52.4(12)	3.0(10)	-56.4(6)	57.0(3)
4	5	-15.2(6)	81.3(5)	-66.0(9)	-8.2(10)	73.9(11)	-58.6(7)
5	4	24.6(6)	52.3(5)	-58.2(4)	-14.3(5)	85.9(5)	-92.9(5)
(c) Chairs							
6	51	-50.8(6)	54.7(3)	-57.9(4)	58.2(4)	-54.3(6)	50.1(7)
7	4	-58.2(4)	78.8(4)	-81.7(7)	81.2(3)	-69.9(3)	52.3(5)
(d) Half-Chairs							
8	26	18.3(6)	1.0(4)	10.9(6)	-42.1(7)	61.7(7)	-48.2(7)
(e) Sofas							
9	4	52.5(12)	-25.3(12)	1.4(7)	0.1(9)	27.9(12)	-54.4(19)
(f) Envelopes							
10	3	-36.1(14)	51.5(9)	-32.9(8)	-2.4(13)	17.7(8)	2.4(8)

Table 4. Superposition of norbornane boat conformations (Table 3, cluster 2) generated by symmetry-modified single-linkage algorithm. Here ±s is the enantiomorph/symmetry permutation applied to the raw torsional sequences (Table 1) to achieve the alignment shown.

Fragment	±s	τ_1	τ_2	τ_3	τ_4	τ_5	τ_6
49	8	0.0	71.4	-71.5	0.0	71.5	-71.4
59	-4	3.8	67.7	-72.3	1.9	70.0	-73.0
63	4	2.2	70.2	-72.0	1.7	70.7	-73.6
65	-12	1.2	72.6	-73.4	0.2	72.3	-73.0
66	1	1.2	72.6	-73.4	0.2	72.3	-73.0
67	4	0.5	70.1	-70.2	-0.3	70.2	-70.4
68	-12	1.6	69.0	-70.2	0.0	70.1	-71.4
69	9	1.7	71.3	-71.7	-1.2	70.4	-70.8
74	4	1.7	69.6	-71.7	-0.1	72.6	-72.3
79	-9	8.3	65.7	-67.1	-6.4	77.3	-80.1
86	12	8.1	66.0	-71.8	4.9	68.2	-75.5
87	12	5.9	64.9	-70.1	2.4	72.6	-76.7
91	12	9.0	63.7	-68.8	-1.1	74.5	-80.6
95	12	2.8	67.8	-70.7	1.1	72.0	-74.9
99	12	2.6	68.1	-70.3	1.0	72.2	-74.5
104	9	7.6	66.9	-68.6	-5.3	78.1	-80.6
114	7	5.2	64.4	-69.8	0.2	72.9	-78.1
121	-5	4.7	65.8	-70.1	1.0	72.5	-72.8
122	9	0.2	74.8	-74.3	-0.6	74.5	-74.6
123	-4	0.4	74.6	-73.6	-0.1	74.5	-74.9
128	4	3.0	69.1	-71.1	1.9	70.0	-73.7
130	-1	1.0	70.0	-68.1	0.6	67.1	-71.6
131	-8	2.5	68.7	-71.1	0.6	70.5	-72.7
132	12	7.9	63.2	-63.2	-7.9	76.9	-76.9
134	6	4.7	68.2	-73.2	1.8	70.9	-75.6
136	-6	4.2	68.7	-72.8	1.5	69.2	-73.3
137	-6	2.9	69.2	-71.9	1.6	70.4	-74.0
139	-3	2.4	69.4	-66.6	-0.4	66.3	-71.3
167	-10	0.4	71.0	-68.3	0.4	68.9	-73.4
168	3	0.9	68.0	-63.5	-2.2	67.2	-69.8
219	-12	0.2	62.0	-60.2	-3.0	65.0	-63.0
220	-2	0.6	66.6	-63.1	-2.4	66.7	-69.1
221	4	-0.5	63.3	-62.4	-1.6	64.4	-63.2
222	11	-0.6	68.8	-63.8	-1.7	66.8	-68.3
Nobs		34	34	34	34	34	34
Mean		2.9	68.3	-69.5	-0.3	70.9	-73.2
Max.		9.0	74.8	-60.2	4.9	78.1	-63.0
Min.		-0.6	62.0	-74.3	-7.9	64.4	-80.6
ESD Sample		2.8	3.1	3.7	2.5	3.3	3.9
ESD Mean		0.5	0.5	0.6	0.4	0.6	0.7

Results from the symmetry-modified single-linkage algorithm are presented in Tables 3 and 4. Step 170 (of 221) was indicated as representing optimum clustering, at which point 177 fragments had been assigned to 10 clusters with $Np \geq 3$. Table 3 shows that the algorithm has successfully broken down the dataset into chemically sensible conformational subdivisions. Indeed, the classification is quite finely detailed, viz : (a) highly puckered norbornane boats (cluster 2) are separated from 'normal' boats (cluster 3) and from the small groups of distorted boats (clusters 4,5); (b) normal chairs (cluster 6) are separated from highly distorted examples (cluster 7). Table 4 shows a section of the torsion angle listing for cluster 2, illustrating the best superposition of conformations achieved by the symmetry-modified algorithm.

Presentation of Results for Use in Molecular Modelling

An automated system for conformational analysis from large subsets of crystallographic data must present results in numerical and (particularly) graphical forms for rapid assesment in a modelling environment. An important visual representation of the symmetry-modified results described above is provided by a principal component analysis of the final cluster at step Nf-1 of the single-linkage method. The continuous re-orientation of fragments means that this final cluster now represents the best superposition of all fragments. Not only is the optimum torsional overlap preserved within clusters, but the clusters are also orientated so that the distances between them tend to be minimised in conformational space. The final cluster therefore represents a unique asymmetric unit of conformational space. The principal component plot of Figure 4 is generated from the sample dataset. The major clusters may be highlighted in different colours, or by a rapid visual 'contouring' as shown here.

A simple statistical summary is generated for each cluster (see Table 4) and results for all clusters are summarised at the end of an analysis, ordered by decreasing cluster size. Cluster means are regenerated at this stage, by taking the original mean as the new cluster 'root', rather than that fragment chosen at random by the single-linkage algorithm. Atomic coordinates are generated for each alternative conformation by locating the 'most representative fragment' in each discrete cluster. This is the fragment which has the torsional sequence most similar [lowest D(pq)] to the mean values for the cluster. This process is illustrated in Table 5. The coordinates can, of course, be used to generate 3D plots of the most representative fragment in each major cluster for visual inspection at the terminal.

Principal Component Analysis of Symmetry-Expanded Data

The methods described above represent techniques for symmetry reduction of the raw torsion angle table generated by GSTAT. The principal component plots for the raw sample (Figure 3) show only inexact symmetry, but can readily be recognised as partial mappings of the full conformational space for six-membered rings (see e.g. Pickett et al., 1970; Cremer et al., 1975) and illustrated in Figure 5. We can intoduce exact symmetry into the principal component plots by expanding the initial torsional sequence for each fragment to

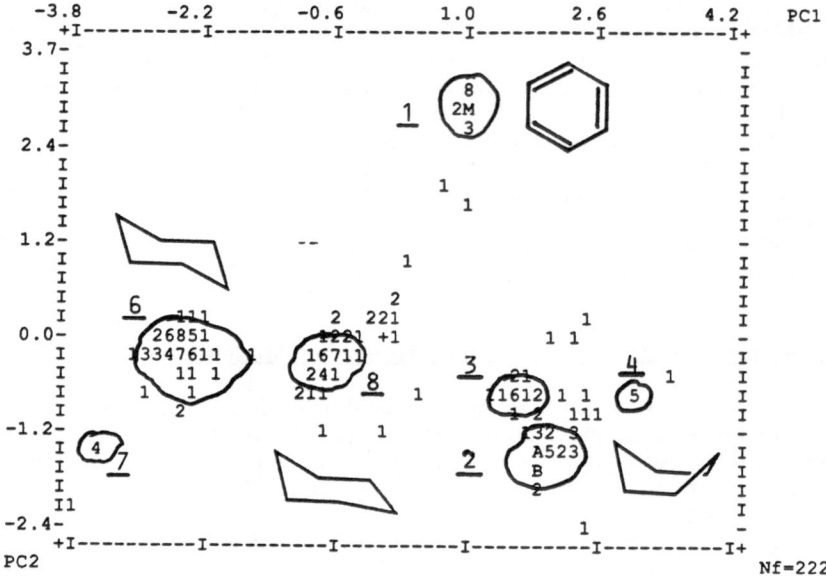

222

Figure 4. Principal component plot for six-membered rings after symmetry-modified clustering. Clusters resolved in this projection are identified by cluster number (underlined) from Table 3.

Table 5. Section from the ranked summary of conformational clustering. Mean values have been re-averaged (see text) by comparison with results in Table 3.

RANK= 1   CLUSTER NO.= 6 POPULATION= 51

Mean values	-55.1	58.6	-57.9	53.3	-50.0	51.1
ESD of Means	0.4	0.4	0.4	0.5	0.7	0.7

Most representative fragment (at D= 5.6 deg from mean) is FRAGMENT NO. 20

MRF Data	-55.3	57.5	-56.3	52.8	-48.5	50.5

Coordinate data for most representative fragment:

ACAMYA	**FRAG**	20**CLUS**	4**RANK**	1
C17	7.55081	3.55694	-7.00195	
C18	6.06258	3.45253	-6.62817	
C19	5.36696	2.15941	-7.16700	
C20	5.60514	2.02631	-8.68815	
C21	7.09969	2.02975	-9.01484	
C22	7.74325	3.28042	-8.49906	

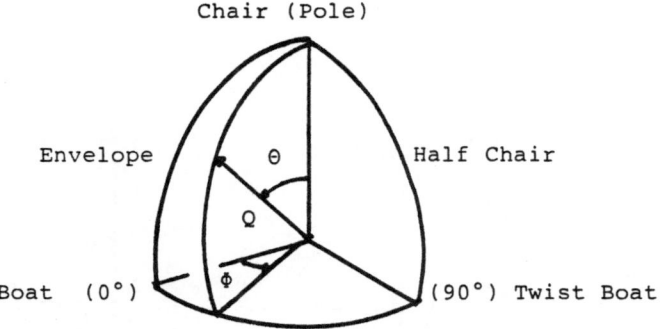

Figure 5. One octant of the spherical conformational space for cyclohexane

indicate all permutations (and their enantiomorphs) of that sequence, as dictated by topological symmetry. For the sample data of six-membered rings, the symmetry-expanded dataset contains 24 x 222 = 5328 torsional sequences. Analysis of the expanded dataset yields three principal components which account for 99.9% of variance (PC1 = 47.9%, PC2 = PC3 = 25.9%). The 2D scattergrams of Figure 6a,b represents projections of all data along two of these three mutually orthogonal principal component axes. Both show exact symmetry and should be compared with their inexact analogues in Figures 3a,b.

Both of the scatterplots of Figure 6 can be readily correlated with the spherical polar description of conformational space for cyclohexane illustrated in Figure 5. Thus Figure 6b represents a projection of this conformational space onto the equatorial plane which has the boat-twist boat-boat pseudorotation itinerary as its equator. This plane is perpendicular to the line connecting the poles (N = chairs, S = enantiomorphic chairs) and passing through the origin (planar phenyl rings). These three major conformations are superimposed in Fig 6b giving rise to the large origin peak. The equator is populated by six large peaks (boats) corresponding to ϕ= 0, 60......300° connected by the very small number of twist-boat distorted conformations (ϕ= 30, 90°,etc.) in this dataset. The central hexagonal ring of density surrounding the origin is generated by the dominant intermediate conformation, the half-chair. The apices of this hexagon point towards the twist-boat areas of the equator as expected. The population density leading from the boat 'peaks' towards the origin are due to the envelope conformation, intermediate between a boat and a chair. The scatterplot of Figure 6a although not so clearly resolved, may be interpreted in a similar manner. The phenyl rings occupy the origin with the ±chairs (poles of the sphere) to the left and right at PC2 = 0. The few highly puckered chairs are clearly distinguished. The vertical line of high population density at PC1 = 0 represents an edge-on view of the boat-twist boat-boat pseudorotation itinerary described above.

The description of conformational space for cyclohexane generated by the principal component method corresponds exactly to that given by Cremer and Pople (1975) from their derivation of generalised ring puckering parameters. Again this is essentially a dimension-reduction process, using the atomic coordinates for each ring as the raw input data. For a general six-membered

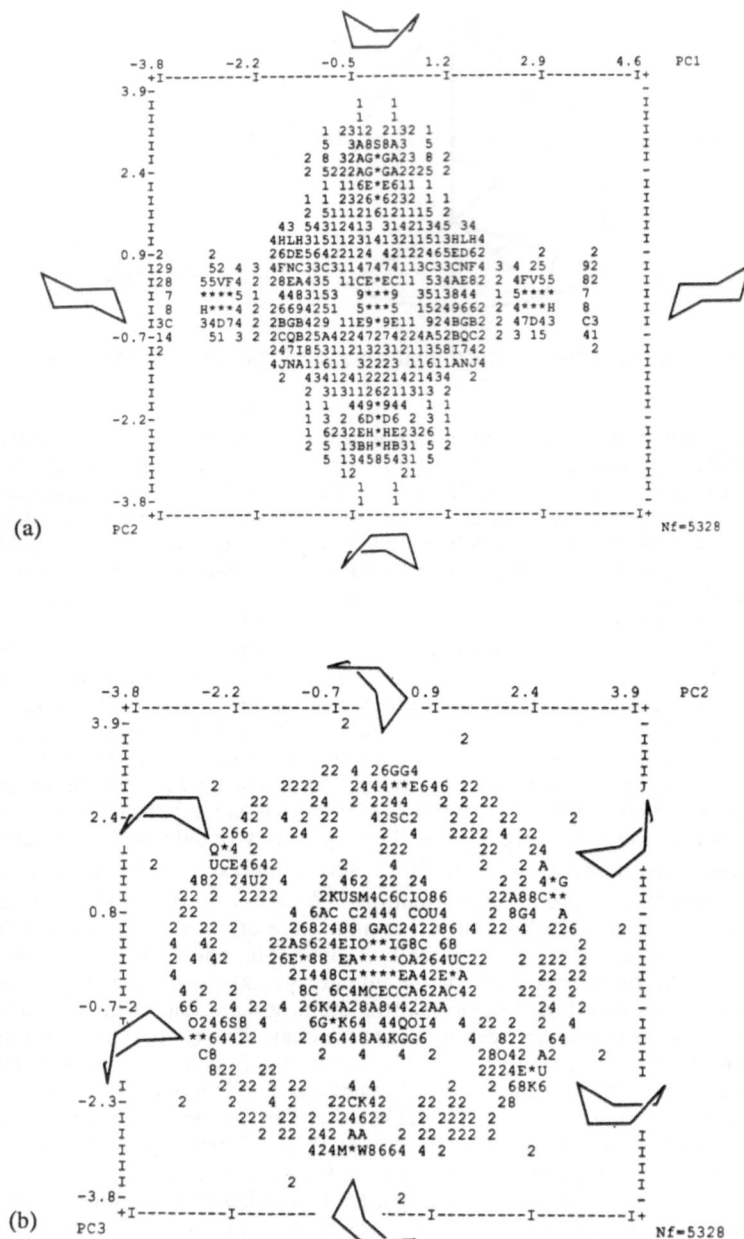

Figure 6. Principal component plots derived from the symmetry-expanded torsional data set for six membered carbocycles.

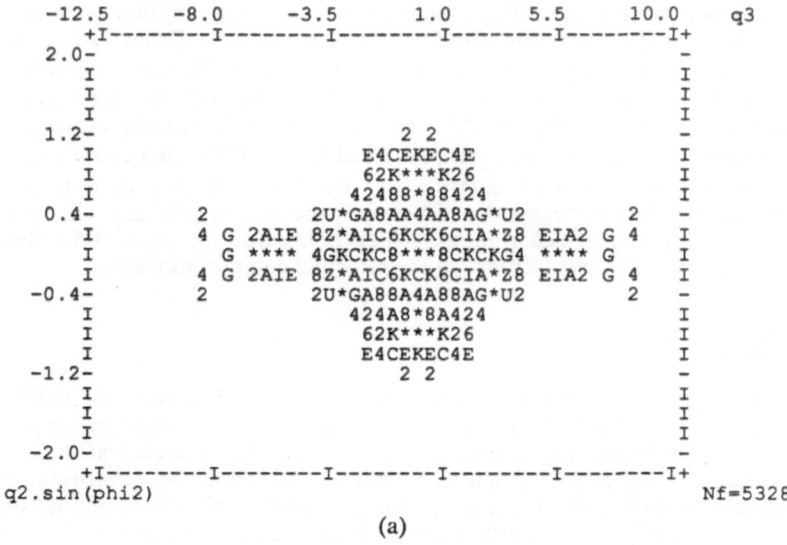

(a)

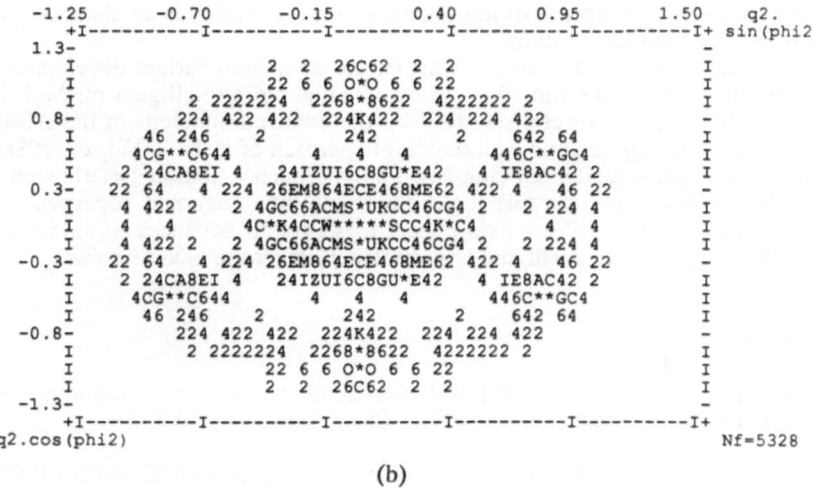

(b)

Figure 7. Scattergrams of Cremer-Pople (1975) puckering parameters derived from symmetry variants of the atomic coordinates of six-membered carbocycles.

ring there are three Cremer-Pople parameters : a single amplitude-phase pair (q_2, ϕ_2) and a single puckering coordinate (q_3), where $q_2 = Q\sin\phi_2$ and $q_3 = Q\cos\phi_2$ in the spherical polar nomenclature of Figure 5. The direct correlation between the principal component axes and the Cremer-Pople description is graphically illustrated in Figure 7a,b. These are normal 2D scatterplots of (a) q_3 vs. $q_2\sin\phi_2$ and (b) $q_2\sin\phi_2$ vs $q_2\cos\phi_2$ derived for all of the ring conformations in the symmetry-expanded dataset (i.e. by atomic rather than torsional permutations). This correspondence between the statistically and chemically derived conformational descriptions, which holds true for other n-membered rings, lends considerable credence to the principal component method.

Conclusion

The large volume of well resolved crystallographic data on small molecules contains a wealth of information, both intramolecular and intermolecular, which is vital to the modelling process. This Chapter has concentrated on the intramolecular aspects, particularly on the detection of conformational preferences, and on methods for the graphical mapping of conformational space and for the study of conformational interconversion pathways. Whilst a knowledge of typical bond lengths and valence angles is important, these parameters tend to span narrow, well known ranges and can be simply displayed as univariate distributions. The detection of conformational preferences is, perhaps, an order of magnitude more difficult, but it is these preferences which dictate the overall 3D shapes of molecules, knowledge which is an absolute pre-requisite in any modelling study.

The methods described in this Chapter are capable of further development. They do, however, offer hope for more automatic, more intelligent methods of model building in the future, presaged by the concepts inherent in the AIMB (analogy and intelligence in model building) approach of Hahn & Wipke (1988). Methods for protein model building based upon sequence homology (Blundell et al., 1987) follow a comparable logic, involving similarity and superposition methods as their basis. We would expect to see rapid advances in automated model building for both small and large molecules over the next few years.

References

Allen F H, Kennard O, Taylor R (1984) Systematic analysis of structural data as a research tool in organic chemistry. Acc. Chem Res. , 16:146-153

Allen F H, Kennard O, Watson D G, Brammer L, Orpen A G, Taylor R (1987). Tables of bond lengths determined by X-ray and neutron diffraction. Part 1. Bond lengths in organic compounds . J. Chem. Soc. Perkin Trans 2, pp S1-S19.

Allen F H, Davies J E (1988). File structures and search strategies for the Cambridge Structural Database. Crystallographic Computing 4 (Ed. Isaacs N W, Taylor M R), Oxford Unversity Press, pp 271-289.

Blundell T L, Sibanda B L, Sternberg M J E, Thornton J M (1987). Knowledge-based prediction of protein structure and the design of novel molecules. Nature, 326: 347-352.

Cremer D, Pople J A (1975). A general definition of ring puckering coordinates. J. Amer. Chem. Soc. , 97: 1354-1358.

Everitt B (1980). Cluster Analysis, 2nd Edition. Halstead Heinemann, Londen.

Hahn M A, Wipke W T (1986). Analogy and intelligence in model building, in Chemical Structures (Ed. Warr W A) Springer-Verlag, pp 269-278.

Jarvis R A, Patrick E A (1973). Clustering using a similarity measure based on shared nearest neighbours. IEEE Trans. Computing, C22: 1025-1034

Murray-Rust P, Bland R (1978a). Computer retrieval and analysis of molecular geometry. II. Variance and its interpretation. Acta Crystallographica, B34: 2527-2533.

Murray-Rust P, Motherwell S (19786). Computer retrieval and analysis of molecular geometry. III. Geometry of the β-1'-aminofuranoside fragment. Acta Crystallographica, B34: 2534-2546

Murray-Rust P (1982). Computer analysis of molecular geometry. V. Symmetry aspects of factor analysis. Acta Crystallographica, B38: 2765-2771.

Murray-Rust P, Raftery J (1985). Computer analysis of molecular geometry. VI. Classification of differences in conformation. J. Mol. Graphics 3: 50-59.

Norskov-Lauritsen L, Burgi H B (1985). Cluster analysis of periodic distributions; application to conformational analysis. J. Computational Chem. , 6: 216-228.

Orpen A G, Brammer L, Allen F H, Kennard O, Watson D G, Taylor R (1989). Tables of bond lengths determined by X-ray and neutron diffraction. Part 2. Organometallic compounds and coordination complexes of the d- and f- block metals. J. Chem. Soc. Dalton Trans, in the Press.

Pickett H M, Strauss H L (1970). Conformational structure, energy and inversion rates of cyclohexane and some related oxanes. J. Amer. Chem. Soc. , 92 : 7281-7290.

Taylor R, Kennard O (1984). Hydrogen-bond geometry in organic crystals. Acc. Chem. Res. , 17: 320-326.

Taylor R (1986a). The Cambridge Structural Database in molecular graphics, techniques for the rapid identification of conformational minima. J. Mol. Graphics 4: 123-131.

Taylor R (1986b). CAMAL - A new component of the Cambridge Structural Database software system. J. Applied Crystallography, 19: 90-91.

Willett P, Winterman V, Bawden D (1986a). Implementation of nearest-neighbour searching in an online chemical structure search system. J. Chem. Inf. Comput. Sci. , 26: 36-41.

Willett P, Winterman V, Bawden D (1986b). Implementation of nonhierarchic cluster analysis methods in chemical information systems: selection of compounds for biological testing and clustering of substructure search output. J. Chem. Inf. Comput. Sci. , 26: 109-118.

Preferred Interaction Patterns
from Crystallographic Databases

R. Scott Rowland, Frank H. Allen,
W. Michael Carson, and Charles E. Bugg

Introduction

A knowledge of three-dimensional structure, in all of its aspects, is an
essential prerequisite of the molecular modelling process. This knowledge may
be divided, on energetic grounds, into two categories. Firstly, information is
required about the covalent aspects of three dimensional structure – bond
lengths, valence angles, and conformational data which dictate the overall
molecular shape. Secondly, geometrical descriptions are needed of the much
weaker interactions by which atoms and molecules associate with each other
in a non-bonded sense. Crystallography is unique in its ability to provide
direct experimental results in both of these areas. The technique is now
being applied to molecules of ever-increasing size and complexity and in ever-
increasing numbers. Details of well over 100,000 crystal structures have been
published – some 400 proteins and biological macromolecules, 76,000 small
molecules containing organic carbon, and nearly 40,000 inorganic, mineral
and metal structures. All of this information is of immense value and the
advent of crystallographic databases makes the data more readily available
in an organized form. It is now a relatively simple matter to locate relevant
structures and extract their coordinates for use in modelling studies.

Despite the success of the crystallographic method, the availability of three-
dimensional structures is low by comparison with the 10 million or so known
chemical compounds. Hence the available three-dimensional data are now
being used to derive systematic information that enables predictions of unknown three-dimensional structures to be made and used in modelling with
some degree of confidence. Crystal structure analysis is itself a modelling
process; crystallographers have always used their own results in this way.
Thus, all structures are judged and discussed against benchmarks of acceptable geometry. Protein structures, in particular, are built and refined using
structural knowledge from both small and large molecules. Over the past
decade there have been considerable advances in the methodology for systematic studies of intramolecular geometry. The ultimate aim is to generate

229

software systems to provide maximum intelligent assistance in model building for unknown three-dimensional structures (a number of these developments are noted in another Chapter of this Volume.)

The present Chapter is concerned with the second aspect of structural systematics identified above – the study of non-bonded interactions and their role in molecular association. This aspect may itself be subdivided, again on energetic grounds, into hydrogen bonds and purely non-bonded interactions which lie at the very heart of the molecular recognition process. Here, the crystallographic method is the *only* experimental technique by which the geometry of these interactions may be studied in a systematic manner. Most of our knowledge about the dimensions and directionality of hydrogen bonds has been derived from crystal structures, and numerous monographs and research articles were published in the 1960's and early 1970's (see e.g., Pimental and McClellan, 1960; Hamilton and Ibers, 1968). The study of limiting contact distances between non-bonded atoms was begun by Pauling in the 1930's (Pauling, 1939), again using crystallographic data as a basis.

There is now a resurgence of interest in these areas, brought about to some degree by the availability of the crystallographic databases. Because non-bonded interactions are energetically weak, it is not possible to detect trends in dimensions or directional preferences from a study of a very small number of crystal structures. Only by rigorous statistical analysis of all available data can convincing evidence be obtained about the nature of these essentially weak interactions. A clear example of these comments is to be found in the history of the $C\text{-}H\cdots O$ hydrogen bond. The importance of $C\text{-}H\cdots O$ interactions in biological systems was first proposed by Sutor (1962) from a study of eight crystal structures. In 1968, Donohue re-examined the evidence available at that time in response to the question "the $C\text{-}H\cdots O$ hydrogen bond – what is it?". He concluded that the answer (at that stage) was "It isn't." Continuing spectroscopic evidence in favor of the interaction prompted a thorough statistical review (Taylor and Kennard, 1982) using 661 $C\text{-}H\cdots X$ interactions from accurate crystal structures. This time the conclusion was that "the majority of short $C\text{-}H\cdots O$ contacts may reasonably be regarded as hydrogen bonds".

A brief overview of crystallographic databases will be our starting point. For the databases utilized in this Chapter, a brief description of their contents will be given. The available computer software will also be discussed since it is crucial for efficiently accessing such databases and calculating geometrical features. Much of the software is in the development stage since the methodology of studying non-bonded interactions is itself still developing. From this developing field two techniques will be described – analysis of preferential association and directionality. Finally, some examples of studies of non-bonded interaction patterns done by other workers and the authors will illustrate some of these emerging techniques.

Description of Crystallographic Databases

The two crystallographic databases that will be discussed in this Chapter are the Cambridge Structural Database (CSD) and the Brookhaven Protein Database (PDB). Most of the crystallographically determined structures of biological interest (such as organic molecules, nucleic acids, and proteins) are stored in these two databases. A brief overview of the more salient features of CSD and PDB is given below. For a more detailed description of these and other crystallographic databases a comprehensive review of the available databases can be consulted (Allen *et al.*, 1987a).

Cambridge Structural Database

The current 1989 version of the Cambridge Structural Database (Allen *et al.*, 1979) contains coordinates for approximately 76,000 small organic molecule structures as determined by X-ray and neutron diffraction analysis. Each structure entry is derived from published coordinates of a diffraction analysis reported in one of the more than 700 journals cited in the CSD. Before inclusion into the database, each structure must pass rigorous tests for errors and checks for internal consistency.

In addition to the coordinate file, an extensive suite of programs is provided for searching the database and calculating molecular geometry. A wide range of parameters (e.g., R-factor, unit cell parameters, presence of a particular element) can be used in combination to search the database for crystal structures of interest. The ability to locate user-defined chemical substructures is one of the most powerful features of the search program. Once a set of structures is located, the geometry analysis program (GSTAT, see e.g. Murray-Rust and Motherwell, 1978 and Murray-Rust and Raftery, 1985) can be used to calculate a variety of geometrical parameters (e.g., bond lengths, sugar pucker, least squares planes, plus user–defined geometry). GSTAT also has options for performing a number of statistical procedures, making it a versatile stand-alone tool.

A considerable amount of information about non-bonded geometry can also be obtained very rapidly via GSTAT. This program has the ability to extend the connectivity basis for three-dimensional substructure searches to include symmetry-related molecules according to user-specified non-bonded distance criteria. Fragments involving non-bonded contacts, within these distance limitations, can now be located and systematic analyses of the fragment geometry performed. A number of special features were added to the 1989 release of GSTAT to facilitate studies of hydrogen bonded systems A$\cdots$H-D (A=acceptor, D=donor) – (a) the positions of H atoms in crystal structures can now be normalized, such that the X-H distance along the bonded vector corresponds to the mean X-H bond length obtained from neutron studies (Allen *et al.*, 1987b) - this is a much more realistic position for the proton; (b) the program can now distinguish between intramolecular and intermolecular

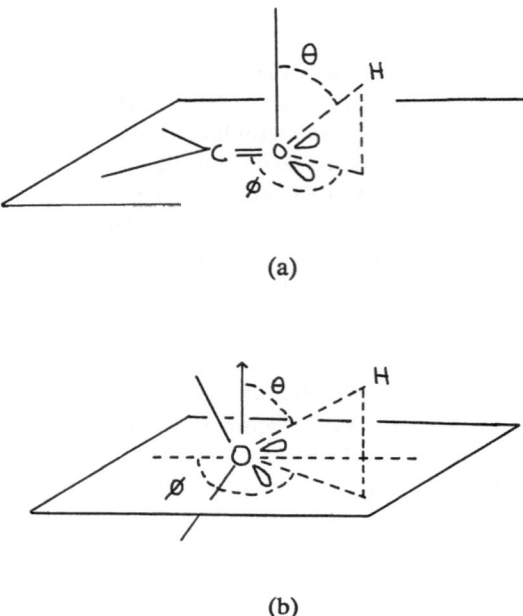

(a)

(b)

Figure 1: The geometric analysis of hydrogen bonded systems using spherical polar angles. (a) sp^2 hybridization geometry. (b) sp^3 hybridization geometry.

non-bonded contacts, including or excluding either category at the discretion of the user; (c) the spherical polar angles θ, ϕ of Figure 1 can be calculated; these angles define the directionality of the A$\cdots$H interaction with respect to either sp^2 or sp^3 lone pairs on A; (d) histograms of the A$\cdots$H-D angles can be corrected for the conical geometric factor (Kroon et al., 1975). Some of these new facilities are also of use in studies of purely non-bonded interactions.

Brookhaven Protein Database

The Brookhaven Protein Database (Bernstein et al. , 1977) is the depository of atomic coordinates for biological macromolecules including proteins, nucleic acids, and polysaccharides. Although the majority of structures were experimentally determined by diffraction analysis, the database does contain some model structures. The current 1989 release of the PDB contains approximately 420 entries, including complexes and homologous proteins from different sources. A large body of user-written software for analysis and graphical display of proteins is generally available to supplement the software distributed with the PDB. One of the authors has modified GSTAT to read in protein structures allowing proteins to be analyzed with the same flexibility as small molecules from CSD.

Analysis of Non-bonded Interactions

Although much can be done in the systematic study of intramolecular properties with available software (e.g., GSTAT), the study of intermolecular properties is a bit more difficult. A certain amount can be done with the available software but there is still much to learn about how to perform these studies, and which mathematical and graphical methods are most useful. Hence, the study of non-bonded interactions is very much a developing research area.

The reason for this apparent subdivision is energetic. In the intramolecular case we deal with accepted chemical frameworks with well-defined 'bonds' and molecular shapes. These are energetically 'hard' parameters by comparison with the very 'soft' interactions in the non-bonded area. In the non-bonded area, therefore, we are often unsure of what we are looking for (at least not in the definite terms of 'bonds', 'angles', etc.). Examples of some developmental approaches for the study of non-bonded interactions will be presented in a later section.

Preferential Association

One of the most basic pieces of information about non-bonded interactions is an estimation of the preferential association of one group of atoms with another. Preferential association of a chemical group is the result of making energetically more favorable non-bonded contacts with one group over another. An estimation of preferences may be obtained by constructing frequency tables of the various types of non-bonded interactions. Such tables usually have to be normalized to correct for the differential number of groups present in the sample or the differing size of the groups. In addition, they must also be tested against a uniform distribution at some level of statistical significance to determine if there is a true preference.

Directionality of Association

A geometric analysis of the non-bonded interaction can be performed to determine if there is any directionality. Directionality occurs when atoms do not interact in a spherically symmetric fashion. An example is the observed difference in the preferred directions of approach of electrophiles and nucleophiles around divalent sulfur (Rosenfield et al., 1977) which will be discussed in more detail in a later section. Analysis of the distribution of contacts around expected lone-pair positions is a common feature of many studies, especially in hydrogen bonded systems. Methods of correcting for geometrical distortion (e.g., the conical distortion in hydrogen bonding) and tests for statistical significance are areas of rapid development.

Patterns of Non-bonded Interactions

Brookhaven Protein Database

Considerable information about amino acid interaction patterns can be obtained from a tabulation of the preferential association of the 20 amino acids using the PDB. Although tables of preferential association have been compiled by other workers (Narayana and Argos, 1984; Warme and Morgan, 1978a; Warme and Morgan, 1978b), the method that we use is unique in that association is not based on contact frequencies but instead on shared surface area between non-bonded neighbors. This method has several advantages over those based on contact frequencies (i.e., strict distance cutoffs). If a distance cutoff between centroids of amino acids is used to define a contact, distortions occur because amino acids are not spherical in shape. The next level of sophistication, atom to atom distance cutoffs, also presents difficulties. The size of the residue becomes a factor since the number of contacts is dependent on surface area (Warme and Morgan, 1978a). By using shared surface area as a definition for association, a degree of physical meaning is given to the tabulated values. Hydrophobic interactions are thought to play an important role in the folding and stability of proteins and in the binding of small ligands to protein sites (Kauzmann, 1959). Although hydrophobicity is a difficult interaction to quantify and model, Chothia (1974) demonstrated that there is an approximate linear relationship between surface area and the free energy of transfer of amino acid residues from an aqueous to a non-polar environment. By using shared surface areas for defining association preferences, a measure of hydrophobicity is implicitly included to some degree.

Surface areas are calculated by using van der Waals dot surfaces constructed around each amino acid residue. The surface is constructed by treating each atom as a sphere with the appropriate radius (see Table 1). Since hydrogens were not considered, the "united atom" concept was used by slightly increasing the van der Waals radii. A 42 point polyhedron is then mapped onto the sphere to generate the dots that comprise the surface. The surface area of a given region on an atom can then be approximated by the number of dots in the region. Calculation of surface area by this method has been shown (Frömmel, 1984) to closely approximate the values calculated by the method of Lee and Richards (1977). The percentage of buried surface area calculated by this method is virtually identical to the values reported by Rose et al. (1985).

Once the dot surface is constructed for the protein chain, each dot is assigned to the nearest non-covalently bonded atom in a neighboring residue or a calculated water molecule. If no atom is within 3.75 Å, the dot is assigned to a water. Inclusion of these "phantom" waters is used to represent the effect of solvent. No crystallographic waters were used in this analysis.

Sums are then tabulated of the number of dots of atom i assigned to atom j in a neighboring group and converted to area. The shared surface area of

Atom	Radius (Å)
O(carbonyl)	1.82
O(hydroxyl)	1.93
N(amide)	1.94
S	2.04
C(carbonyl,trivalent aromatic)	2.06
CH(aromatic)	2.17
C(methyl)	2.18
C(methylene)	2.23
C(alpha)	2.31

Table 1: Radii used to generate van der Waals surfaces of amino acids. Radii were taken from the CEDAR package (Carson and Hermans, 1985) and are based on the equilibrium distance given by Lennard–Jones non–bonded force constants. They are "united atom" radii which are slightly larger than standard van der Waals radii to account for hydrogens when they are not explicitly included.

a side chain can then be calculated by summing the areas of its constituent atoms. One can then define E_{ij} (Lifson and Sander, 1980), the expected shared area of a side chain of type i with type j,

$$E_{ij} = A_i A_j / A_{tot}$$

where A_i and A_j are the sums of the areas of side chains of type i and j, respectively. A_{tot} is the total surface area of all residues. The preferential association or shared surface correlation, A_{ij}, of side chain i for j is defined by,

$$A_{ij} = S_{ij} / E_{ij}$$

where S_{ij} is the observed shared surface of side chain i for j. Stated another way, A_{ij} is the contact correlation of a given group i to be in contact with group j. Normalization with the E_{ij} term also insures that the correlation is not distorted by the difference in surface area of the amino acids. When the contact correlation is favorable, $A_{ij} > 1$, while an unfavorable correlation is indicated by $A_{ij} < 1$. In general, $A_{ij} \neq A_{ji}$ due to the unique microenvironment of each amino acid.

A matrix of the A_{ij}'s for each side chain is shown in Figure 2. Side chains are identified by their standard one letter codes. The peptide backbone atoms are treated separately, except in the case of glycine and proline, and are indicated by B. Cystines and cysteines are treated separately and are represented by U and C, respectively. All heteroatoms present in a protein structure (e.g., ligands, cofactors, ions) are grouped together and are identified by X. Heteroatoms are included for completeness even though the diverse chemical nature of their constituent groups makes interpretation of their A_{ij}

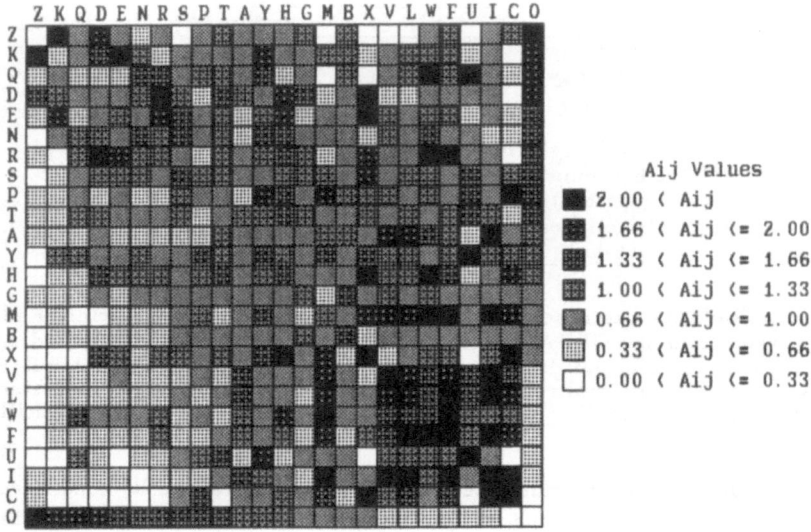

Figure 2: Matrix of amino acid side chain preferential association values, A_{ij}, based on shared surface area. A_{ij} values are represented by shading as shown in the key. Standard one letter codes are used for the side chains. Backbone atoms are indicated by B, heteroatoms X, terminal groups Z, cystine U, cysteine C, and water O. The preference of side chain i for j is found by locating i's row and moving across to the column of j at which point the A_{ij} is read off. Note that in general $A_{ij} \neq A_{ji}$. The rows of amino acids are sorted by hydrophilicity with the most hydrophilic at the top. Hydrophilicity is determined by the A_{ij} with O. Since O-O interactions are not counted the A_{ij} value defaults to zero.

difficult. Terminal amino and carboxyl groups are also treated separately and are indicated by Z. The code O represents the calculated water molecules. A total of 23 proteins was used to construct this matrix of A_{ij}'s. This set of globular monomeric proteins is representative of the well refined structures in the PDB.

Although complex, the matrix of A_{ij}'s contains a wealth of information about the preferential association of amino acid side chains. Only the more salient features of the matrix will be discussed here (a more detailed analysis will be published elsewhere). One of the most striking features is the varying degree of selectivity expressed by the amino acids. The selectivity of an amino acid is reflected by both the range and magnitude of the A_{ij}'s describing its preference for contact with other amino acids. Amino acids with high selectivity (e.g., lysine and phenylalanine) have a wide range of A_{ij}'s indicating very strong tendencies, both positive and negative, toward contact formation. In contrast, there are amino acids whose correlation values are closely grouped together around a value of 1.0 (e.g., alanine and tyrosine). Since a correlation value of 1.0 corresponds to the amount of contact expected by chance, such amino acids have little selectivity.

Close inspection of the more selective side chains reveals two dominating factors: charge and hydrophobicity. Charged amino acids prefer to be in contact with oppositely charged side chains while avoiding hydrophobic groups. Conversely, hydrophobic side chains avoid charged groups and tend to be in contact with other hydrophobic groups. It is worth noting that almost all of the interactions with preference values above 2.0 (having twice the shared surface area expected from a uniform distribution) are hydrophobic. This is in agreement with the observation of Bryant and Amzel (1987) that hydrophobic residues make twice as many hydrophobic contacts than would be expected by chance. This implies that our method of calculating association correlations does an adequate job of modeling hydrophobicity. Other workers (Eisenberg and McLachlan, 1986) have also used surface area to model hydrophobicity.

Among the hydrophobic side chains, the side chain of phenylalanine should be noted in particular as it exhibits several interesting features. First, the preference of phenylalanine for itself is slightly larger than 3.0, one of the highest values in the matrix. This observation is consistent with recent discoveries about the character of aromatic side chains. Several workers (Burley and Petsko, 1985; Singh and Thornton, 1985) have shown that Phe-Phe clustering is the most common type of aromatic-aromatic interaction in proteins. It has been postulated that such interactions may play a role in the stabilization of protein structure. Such interactions have also been shown to have a preferred geometry — the aromatic rings tend to be perpendicular to each other rather than parallel. This "herringbone" pattern is similar to the packing of molecules in the crystal structure of benzene (see Figure 3).

One possible explanation for the observed aromatic–aromatic interaction pattern is in terms of an electrostatic model. Since H atoms have a partial positive charge and C atoms a partial negative charge in aromatic molecules, there is a coulombic attraction that favors close C$\cdots$H approaches and causes the edge-to-face interaction between aromatic molecules. This model has been the basis of suggestions that amino-aromatic contacts would be favorable (Burley and Petsko, 1986) and could possibly explain the high preference of Phe for Arg (> 2.0) in the preference table. However, this observation could be an artifact due to the low occurrence of both Phe and Arg in proteins. In order to further examine the role of electrostatic effects in aromatic interaction patterns, one needs many accurate structures, which makes the CSD an ideal choice.

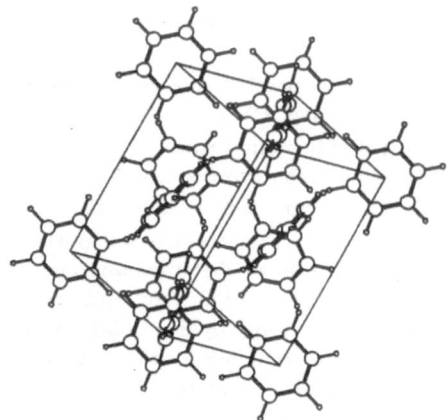

Figure 3: Crystal packing diagram of benzene. Notice the "herringbone" arrangement of the molecules. Coordinates were obtained from the CSD entry BENZEN.

Cambridge Structural Database

The large number of accurate structures combined with the powerful geometry analysis program GSTAT make the CSD an excellent tool for the analysis of non-bonded interactions and their directionality. The effectiveness of the GSTAT approach is illustrated by a typical study involving the $C=S\cdots H\text{-}N$ system. Statistical results for the oxygen analog $C=O\cdots H\text{-}N$ have already been published (Taylor *et al.*, 1983). This work shows that (a) the $O\cdots H\text{-}N$ angle tends towards linearity, and that shorter $O\cdots H$ distances are correlated with straighter hydrogen bonds; (b) there is a distinct preference for the hydrogen bonds to form in, or near to, the directions of the $O sp^2$ lone pairs, i.e., with (θ, ϕ) close to ($90°$, $120°$). It is now a simple matter to survey the available results for the $C=S\cdots H\text{-}N$ system, should such data be required in a particular modelling study. Here a comparison of directionality in the oxygen and sulphur cases would be essential, to assess the effects (if any) of possible d-orbital participation in the case of sulphur.

First we locate the 259 entries in CSD which contain both $C=S$ and N-H bonds and are organic structures, with an R-factor below 0.10, and coordinates available in the database. The $C=S\cdots H\text{-}N$ fragment is constructed with a limit of 3.6Å on $S\cdots N$ and 3.1Å on $S\cdots H$. The fragment is then located in the extended connection table, constructed using the same $S\cdots N$ limit. A variety of geometrical parameters can be tabulated for the fragment and displayed as histograms or scattergrams. Two of these scattergrams, relating to the known results for the oxygen analog, are illustrated in Figure 4. In this study the H atom positions have been normalized to yield an N-H distance

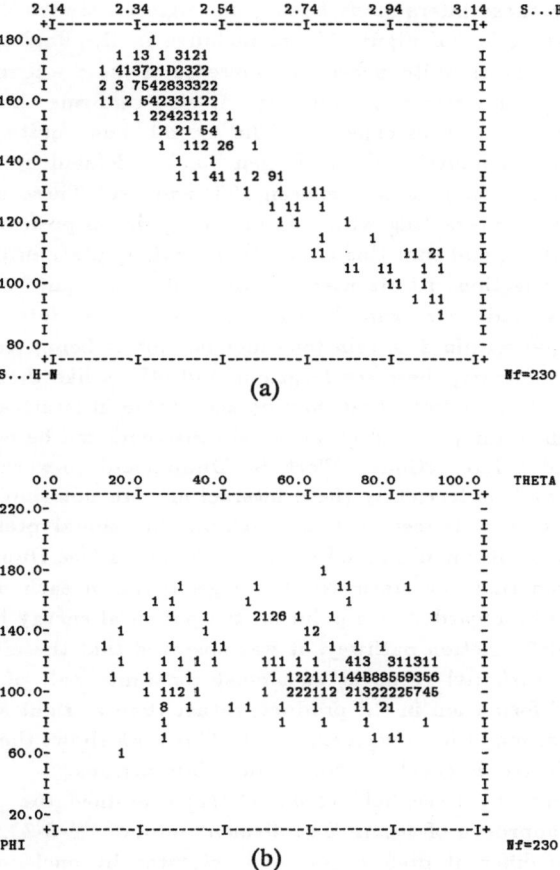

<pre>
       2.14   2.34   2.54   2.74   2.94   3.14   S...H
      +I---------I---------I---------I---------I---------I+
180.0-   1   -
    I   1 13 1 3121   I
    I   1 413721D2322   I
    I   2 3 7542833322   I
160.0- 11 2 542331122   -
    I   1 22423112 1   I
    I   2 21 54 1   I
    I   1 112 26 1   I
140.0-   1   1   -
    I   1 1 41 1 2 91   I
    I   1   1 111   I
    I   1 11 1   I
120.0-   1 1   1   -
    I   1   1   I
    I   11   11 21   I
    I   11 11   1 1   I
100.0-   11 1   -
    I   1 11   I
    I   1   I
    I   I
 80.0-   -
      +I---------I---------I---------I---------I---------I+
 S...H-N   Nf=230
</pre>

(a)

<pre>
       0.0   20.0   40.0   60.0   80.0   100.0   THETA
      +I---------I---------I---------I---------I---------I+
220.0-   -
    I   I
    I   I
    I   I
180.0-   1   -
    I   1   1   11   1   I
    I   1 1   1   1   I
    I   1 1   1   2126 1   1   1   I
140.0-   1   1   12   -
    I   1   1   11   1 1 1 1   I
    I   1   1 1 1 1   111 1 1   413 311311   I
    I   1   1 1 1   1 122111144B88559356   I
100.0-   1 112 1   1   222112 21322225745   -
    I   8 1   1 1   1   1   11 21   I
    I   1   1   1   1 1   1   I
    I   1   1 11   I
 60.0-   1   -
    I   I
    I   I
    I   I
 20.0-   -
      +I---------I---------I---------I---------I---------I+
 PHI   Nf=230
</pre>

(b)

Figure 4: Geometry of S···H-N interaction. Scattergrams were produced by GSTAT which uses the sequence 1-Z to represent the number of observations at a particular point. (a) Plot of the S···H distance (horizontal axis) versus the angle S···H-N (vertical axis). The increase of the S···H separation distance as the S···H-N angle deviates from linearity is apparent. The plot also shows a preference for the linear configuration, as expected. (b) Plot of θ (horizontal axis) versus ϕ (vertical axis). Most of the hydrogens lie at θ angles approaching 90° and ϕ angles between 100°-120°, the direction of the sulfur lone-pairs in sp^3 hybridization.

of 1.009Å. Only those interactions having an angle ϕ between 60° and 180° (in the sp^2 construction of Figure 1) are included in the final results. The first scattergram shows quite clearly the increase of the S$\cdots$H separation as the S$\cdots$H-N angle deviates from linearity. It also confirms that the linear configuration is preferred, as expected. The second shows quite clearly that the preferred directionality of the hydrogen bond is defined by θ angles approaching 90° and with ϕ angles between 100° and 120° These data suggest that the H atom is interacting with sulphur lone pairs in positions expected for sp^2 hybridization, and that this interaction is taking place in the lone-pair plane. The 3D selection criteria used in this study were quite unrestrictive. No decision was made, for example, to regard only those interactions having S$\cdots$H-N angles within a certain tolerance of 180° as being true hydrogen bonds. On the contrary, these scattergrams and others like them should be used to assess the restrictions that may be acceptable in later calculations.

Similar search techniques and geometrical constructs can be used to study purely non-bonded interactions. Work by Dunitz and co-workers through the 1970's revealed quite clearly these interactions are not simply governed by limiting contact distances, but also exhibit directional preferences. A study of intramolecular non-bonded N$\cdots$C=O distances (see Dunitz, 1979 for a review) showed that the static N$\cdots$C=O geometry in each of 14 crystal structures could be regarded as a point on the potential energy hypersurface that maps an SN2 reaction pathway. It was observed that the nitrogen atom approached the carbonyl group at an almost constant angle of 109° , ready for Csp^3-N bond formation in the product, rather than at right angles to the carbonyl plane as might have been expected. This work shows the importance of long-range electronic effects in non-bonded interactions.

A further study by Rosenfield et al. (1977) examined the directions of intermolecular approach of atoms X to divalent sulphur (Y-S-Z). They noted that completely different preferences were exhibited by nucleophilic X and electrophilic X. We have reworked this example using the considerably larger datasets now available for X=Cl or Cl⁻ (nucleophilic) and for X=N⁺ or -NO2 (electrophilic). The geometrical construct of Figure 5 was used, where θ is the angle of declination from the normal to the Y-S-Z plane, and ρ is the maximum of the two equivalent 'valence' angles Y-S$\cdots$X and Z-S$\cdots$X. The θ-ρ scatterplots for the different X, Figure 6, have a similar form. However for the nucleophiles (a) there is a clear preference for θ to approach 90° and ρ to approach 180°; this trend is reversed for the electrophiles (b) where θ peaks between 10°-30° and ρ is about 90°. Thus the nucleophiles prefer to approach S in the Y-S-Z plane and along the extension of the Y-S or Z-S bonds, i.e. in a position to interact with the LUMO of the system, while the electrophiles approach along the direction expected for the lone pair orbital, the HOMO of the system (Rosenfield et al., 1977).

In this survey, we have also examined the intermolecular non-bonded distance S$\cdots$X as a function of θ and show the resultant scatterplots in Figure 7

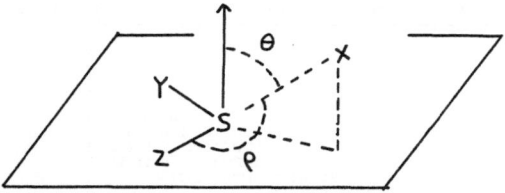

Figure 5: Divalent sulfur interaction geometry.

for X=Cl and X=N, respectively. There is a tendency for the shorter S⋯X distances to be associated with the higher θ angles in both cases, irrespective of their preferred directions of approach. We have divided the datasets into two θ-groups- 0°-40° and 70°-90° - and obtained the mean S⋯X distances shown in Table 2. We stress that these distances are simple means of S⋯X distances out to our limit of 4.0Å in each case. They are not to be taken as closest approach distances (which showed the same trends as the means). Although this study is very preliminary, it does correlate well with results obtained by Nyburg and Faerman (1985) for halogen⋯halogen non-bonded interactions. Here there was clear evidence for increasingly 'anisotropic' van der Waals radii with increasing size of the halogen. The shorter distances were observed for 'head-on' approach of two C-Halogen groups, the longer distances were associated with mutually perpendicular directions of approach. Results of this type suggest that non-bonded atoms are not spherical, and that van der Waals radii are dependent on the chemical environment of the atom and the direction of approach. They may not, therefore, be transferable from one chemical structure to another. Further work must involve systematic studies of non-bonded contacts between well-defined chemical functional groups of the type described below. This will, in turn, reveal standard methodologies for the study of non-bonded systematics which can be used to enhance the publicly available software system.

	0°-40°		70°-90°	
	S⋯X	Nobs	S⋯X	Nobs
N	3.635(14)	64	3.481(34)	24
Cl	3.658(14)	61	3.552(14)	131

Table 2: Mean S⋯X distances for divalent sulfur system. The datasets were divided into two θ ranges, 0°-40° and 70°-90°. For each range mean distances are given for Cl and N. Nobs is the number of observations.

242

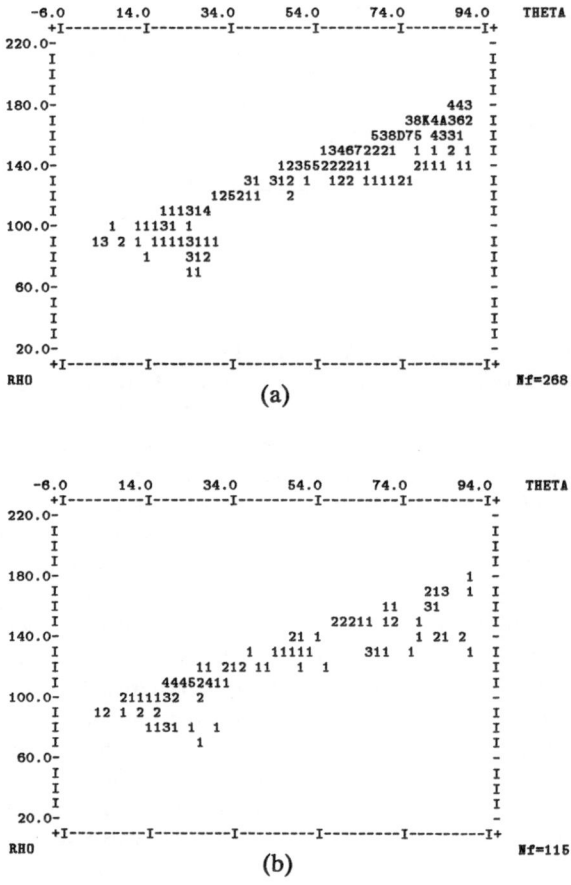

(a)

(b)

Figure 6: Nucleophile and electrophile approach to divalent sulfur. Scattergrams show θ (horizontal axis) plotted versus ρ (vertical axis). (a) Nucleophiles. The preferred direction is close to $\theta = 90°$ and $\rho = 180°$. In this position the nucleophile can interact with the LUMO of the system. (b) Electrophiles. In contrast with the nucleophiles, θ lies between 10°-30° and ρ tends to 90°, the direction of the HOMO of the system.

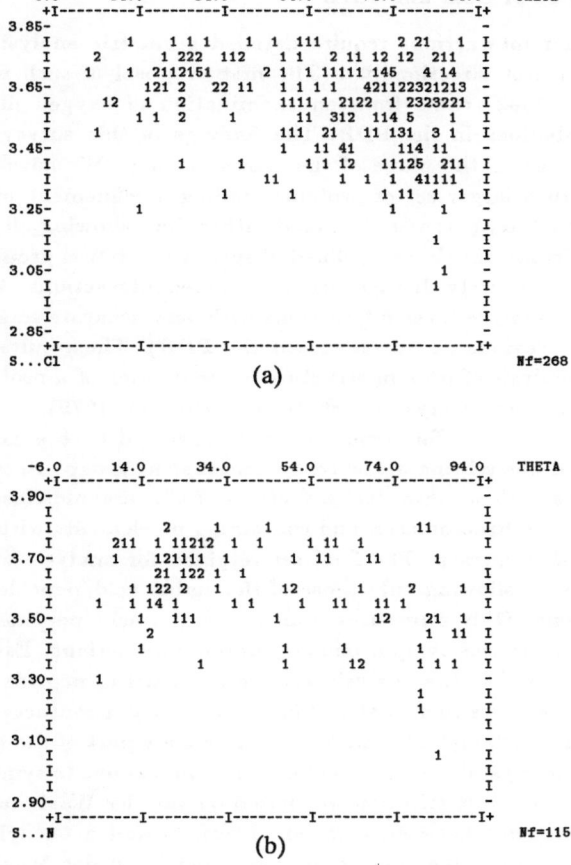

Figure 7: Distance of Cl and N to divalent sulfur as a function of θ. Note the tendency for shorter S···X distances at high θ values for both CL and N. (a) Cl atoms. (b) N atoms.

Combined Use of CSD and PDB

Aromatic-polar interactions require detailed geometric analysis to determine the presence of directionality. The first proposal of such interactions (Thomas et al., 1982) came from an examination of oxygen distributions around phenylalanines in the PDB. The findings of this survey were that oxygens tend to lie in the plane of the aromatic ring. We have duplicated the analysis with a larger set of proteins, adding a refinement in that only backbone carbonyl oxygens were included rather than allowing all oxygens as in the previous study. Such well-defined chemical functional groups must be used in order to accurately characterize non-bonded interactions. We also utilized the CSD to analyze these interactions with very accurate small molecule structures (the previous group only used the PDB). The results should be similar to the analysis of protein structure if the interior of a protein is truly analogous to a molecular crystal (Schulz and Schirmer, 1979).

The CSD was searched for structures that contained both a carbonyl and an unsubstituted phenyl ring attached by a methylene group. Restricting the search to entries with no disorder, R-factor $< 7.5\%$, organic class, intensity data measured by diffractometer, and containing no elements with $Z>18$ produced a set of 785 entries. The final set retained for analysis was reduced to 90 structures by allowing only those of the amino acid/peptide class with located hydrogens. Only structures from the amino acid/peptide class were used since they more closely approximate protein interactions. Essentially all of the carbonyls used in the analysis were components of peptide bonds.

These structures were then analyzed for intermolecular contacts involving a phenyl ring and a carbonyl oxygen. Such contacts are part of the interactions that contribute to crystal packing. Redundant contacts due to symmetry were not allowed. The contact criterion was based on van der Waals radii (Bondi, 1964). If the distance between a carbonyl oxygen and a phenyl carbon or hydrogen was less than the sum of the two atoms' van der Waals radii the two groups were considered to be in contact. Thus, the dataset consists of contacts shorter than optimal van der Waals contacts. Hydrogen positions were normalized to standard bond lengths. A total of 69 carbonyl–phenyl contacts were located. Inclusion of the carbonyl carbon in the contact search produced no significant contacts.

For all of the contacts located, a complete set of geometrical parameters was calculated. A spherical coordinate system was used to describe the location of the carbonyl oxygens with respect to the phenyl ring (see Figure 8). All contacts were reduced to one quadrant by the mm symmetry of the aromatic ring. In addition, hydrogen bonding geometry was calculated with the closest phenyl C–H as the donor as shown in Figure 1a.

A $\theta - \phi$ scattergram of the oxygen position with respect to the aromatic ring is shown in Figure 9a. Approximately 85% of the carbonyl oxygens are in the range $70° < \theta < 90°$ (i.e., are close to lying in the plane of the aromatic

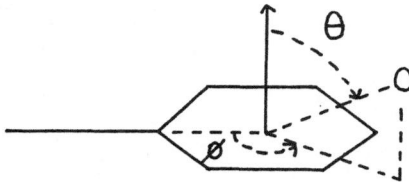

Figure 8: Spherical coordinate system used to measure the geometry of carbonyl oxygens around an aromatic ring.

ring). However, in a spherical coordinate system incremental sections along θ do not have the same volume. For this reason, the θ distribution was tested to determine if it was statistically distinguishable from an uniform distribution given by $\sin\theta$ in a spherical coordinate system (Singh and Thornton, 1985). The probability of the distribution occurring by chance is less than 0.01 as calculated by a χ^2 analysis. The oxygens also tend to lie at ϕ values close to the hydrogen positions of the aromatic ring. Figure 10a illustrates the distribution of carbonyl oxygens around a reference phenyl ring using pseudo-density contours at the 50% and 25% levels for this set (Rosenfield et al., 1984). The clustering of the oxygens around the hydrogen atoms of the phenyl ring is quite apparent.

A closer look at how the aromatic hydrogens are oriented around the carbonyl group is shown in the $\theta - \phi$ scattergram of Figure 9b. Although no strong trend exists, the $\theta - \phi$ values are in regions generally accepted as hydrogen bonds ($35° < \theta < 90°, 120° < \phi < 180°$.) The lack of clustering around the lone-pair makes classification of the interaction as a hydrogen bond tenuous. However, a previous study of various C-H$\cdots$X hydrogen bonds (Taylor and Kennard, 1982) obtained similar results.

Brookhaven Protein Database

The PDB was then searched for similar phenyl–carbonyl contacts. In this case the search was confined to intramolecular contacts. The CSD geometry analysis software, GSTAT, was modified to handle proteins in order to perform this search. A set of 36 proteins whose structures had been determined to 2.0Å resolution or better was used for this analysis. The contact criterion was the same as used for the CSD. This required the calculation of expected hydrogen coordinates for the phenylalanine residues. Hydrogens were calculated to lie in the aromatic plane and have bond lengths corresponding to observed values from neutron diffraction studies (Allen et al., 1987b). In

246

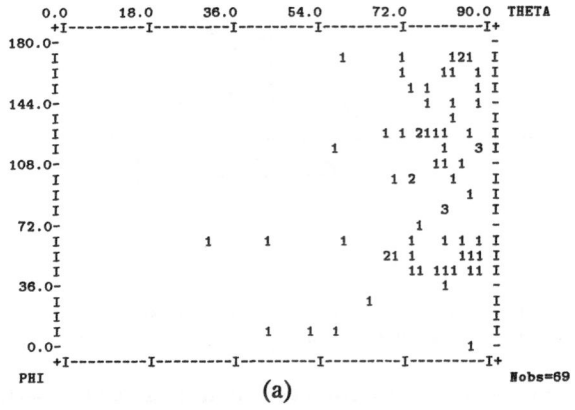

(a)

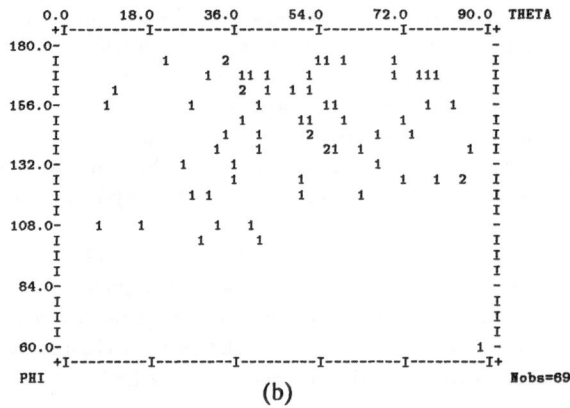

(b)

Figure 9: Aromatic$\cdots$O=C interaction geometry from the CSD. (a) Location of carbonyl oxygens with respect to the aromatic ring. The geometric construct of Figure 7 was used to describe the location of the oxygens. Most of the oxygens are close to the aromatic plane i.e., $70° < \theta < 90°$. (b) Location of the closest C-H hydrogen to the carbonyl group. The geometric construct of Figure 1a was used to describe the location of the hydrogens.

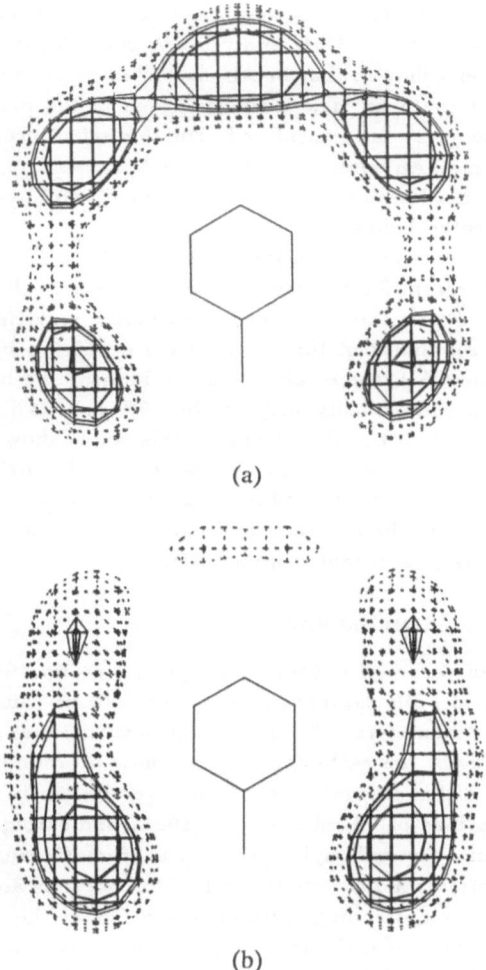

(a)

(b)

Figure 10: Psuedo-density contours illustrating the distribution of carbonyl oxygens around a reference phenyl ring. Each oxygen is represented by an isotropic Gaussian smearing function. A density map is then generated by summing over all the atoms at each point. The dotted line contours contain 50% of the map's total integrated density and the solid line contours 25%. Both maps were generated using the subset of oxygen contacts that are closer than optimal van der Waals contacts and normalized for the number of contacts. (a) Contours for the CSD distribution. (b) Contours for the PDB distribution.

addition, only backbone carbonyls were used so that the results would be directly comparable with the CSD analysis. A subsequent search allowing other types of oxygens showed that backbone carbonyls accounted for 85% of all carbonyl–phenyl contacts. The carbonyl group in the phenylalanine and its adjacent residues were not considered. For this set of proteins, a total of 34 contacts was found. An analysis adding a 1.0Åtolerance to the contact criterion to allow for coordinate error (data not shown) exhibited similar trends to the results presented below.

Analysis of the geometry was carried out as in the CSD. The Figure 11a shows a $\theta - \phi$ scattergram for the contacts with respect to the aromatic ring. The range $70° < \theta < 90°$ contains 75% of the carbonyl oxygens. A χ^2 analysis verified that the probability of this distribution occurring by chance is less than 0.01. As in the CSD, the ϕ values tend to be near the hydrogens of the aromatic ring. A pseudo-density map at the 50% and 25% levels illustrating the distribution of carbonyl oxygens in this set is shown in Figure 9b. Although the predominant clustering occurs around the ortho position, the other hydrogen positions also show clustering. Figure 11b contains a scattergram of the geometry of the aromatic hydrogens around the carbonyl group. It is similar to the results obtained from CSD.

Relationship Between CSD and PDB

Since contacts of carbonyl oxygens with phenyl groups were defined and located identically in both databases, the results can be compared. Similarities in geometric parameters calculated from both databases are apparent. Both θ distributions of the carbonyl oxygens show a pronounced preference to lie in the plane of the aromatic ring. These results are in agreement with the earlier observations of Thomas et al. (1982), indicating that there is a directional preference for carbonyl oxygens near aromatic rings. The pseudo-density maps (Figure 9) indicate that most of the oxygens are clustered near a hydrogen of the aromatic ring. Although analysis of the directionality of the hydrogens around the carbonyl group was not conclusive the complete study suggests that some form of weak hydrogen bonding may be occurring.

The fact that there are so many similarities in the geometry of the aromatic-carbonyl interaction in both databases reinforces the concept of the interior of proteins being like an organic crystal. The most striking difference in the geometry between the two databases is the ϕ distribution of the carbonyl oxygens. In the CSD, the oxygens are evenly distributed around each hydrogen position while in the PDB the ortho position is the predominant location of oxygens. An examination of the contacts in the PDB revealed that most of the aromatic rings were also forming hydrophobic interactions at the other positions leaving only the ortho position available.

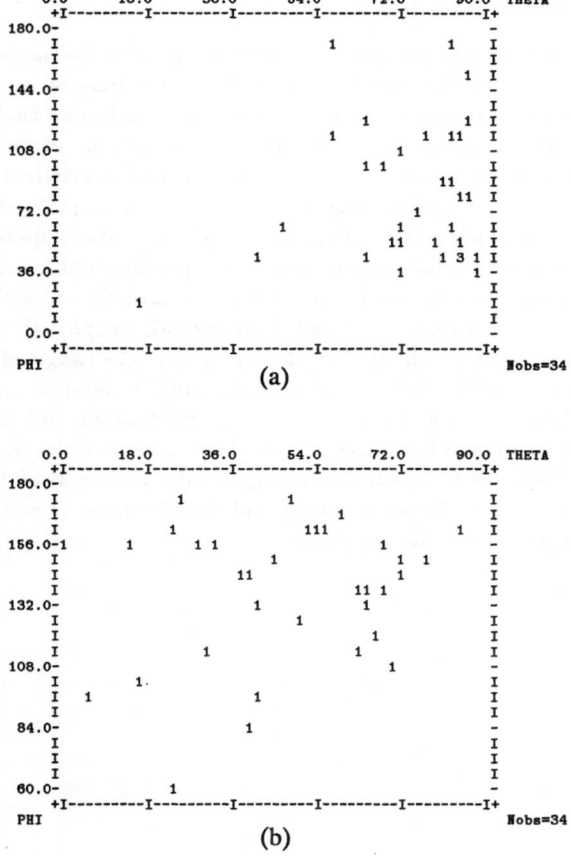

Figure 11: Aromatic···O=C interaction geometry from the PDB. (a) Location of carbonyl oxygens with respect to the aromatic ring. The geometric construct of Figure 7 was used to describe the location of the oxygens. Most of the oxygens are close to the aromatic plane i.e., $70° < \theta < 90°$. (b) Location of the closest C-H hydrogen to the carbonyl group. The geometric construct of Figure 1a was used to describe the location of the hydrogens.

Conclusions

Crystallographic databases are an important tool in the molecular modelling process. They provide rapid access to large numbers of experimentally
determined molecular structures that can serve as standards. Such databases
are frequently used to study the molecular conformations of chemical fragments. The development of computer software and statistical techniques
allow analysis of bond lengths, angles, and other geometrical parameters to
be performed quickly with little effort from the user. Methodologies for the
routine study of non-bonded interactions via crystallographic databases are
only now being developed. Both the preference and directionality of such
interaction patterns can be determined from crystallographic databases. Examples of such studies given in this chapter show the range of techniques
being developed. Results from some of these studies tend to indicate that
atoms do not interact in a spherically symmetric fashion and that van der
Waals radii are dependent on the chemical environment of the atom and
the direction of approach. Such interesting results ensure that the study of
interaction patterns will be an exciting and fruitful area of research in the
utilization of crystallographic databases.

References

Allen, F. H., Bellard, S., Brice, M. D., Cartwright, B. A., Doubleday, A., Higgs, H., Hummelink, T., Hummelink-Peters, B. G., Kennard, O., Motherwell, W. D. S., Rodgers, J. R., Watson, D. G. (1979). The Cambridge Crystallographic Data Centre: computer-based search, retrieval, analysis and display of information. Acta. Crystallogr., Sect. B; Struct. Sci. B35, 2331-2339.

Allen, F. H., Bergerhoff, G., and Sievers, R. (1987a). Crystallographic Databases. Polycrystal Book Service, Dayton, Ohio.

Allen, F. H., Kennard, O., Watson, D. G., Brammer, L., Orpen, A. G., and Taylor, R. (1987b). Tables of bond lengths determined by X-ray and neutron diffraction. Part 1. Bond lengths in organic compounds. J. Chem. Soc. Perkin Trans. II , S1-S19.

Bernstein, F. C., Koetzle, T. F., Williams, G. J. B., Meyer, E. F., Jr., Brice, M. D., Rodgers, J. R., Kennard, O., Shimanouchi, T., and Tasumi, M. The Protein Data Bank: A computer-based archival file for macromolecular structures. (1977). J. Mol. Biol. 112, 535-542.

Bondi, A. (1964). van der Waals volumes and radii. J. Phys. Chem. 68, 441-451.

Bryant, S. H., and Amzel, L. M. (1987). Correctly folded proteins make twice as many hydrophobic contacts. Int. J. Peptide Protein Res. 29, 46-52.

Burley, S. K., and Petsko, G. A. (1985). Aromatic-aromatic interaction: a mechanism of protein structure stabilization. Science 229, 23-28.

Burley, S. K., and Petsko, G. A. (1986). Amino-aromatic interactions in proteins. FEBS 203, 139-143.

Carson, M., and Hermans, J. (1985). Molecular dynamics workshop laboratory. In "Molecular Dynamics and Protein Structure" (J. Hermans,ed.), pp. 165-166. Polycrystal Book Service, Dayton, Ohio.

Chothia, C. H. (1974). Hydrophobic bonding and accessible surface area in proteins. Nature 248, 338-339.

Dunitz, J. (1979) X-ray Analysis and the Structure of Organic Molecules. Cornell University Press, Ithaca.

Eisenberg, D. and McLachlan, A. D. (1986). Solvation energy in protein folding and binding. Nature 319, 199-203.

Frömmel, C. (1984). The apolar surface area of amino acids and its empirical correlation with hydrophobic free energy. J. Theor. Biol. 111, 247-260.

Hamilton, W.C. and Ibers, J.A. (1968). Hydrogen Bonding in Solids. Benjamin,New York.

Kauzmann, W. (1959). Some factors in the interpretation of protein denaturation. Adv. Protein Chem. 14, 1-63.

Kroon, J., Kanters, J.A., van Duijneveldt-van de Rijdt, J.G.C.M., van Duijneveldt, F.B., and Vliegenthart, J.A. (1975). O-H···O hydrogen bonds in molecular crystals: A statistical and quantum-chemical analysis. J. Mol. Struct. 24, 109-129.

Lee, B. and Richards, F. M. (1971). The interpretation of protein structures: estimation of static accessibility. J. Mol. Biol. 55, 379-400.

Lifson, S. and Sander, C. (1980). Specific recognition in the tertiary structure of β-sheets of proteins. J. Mol. Biol. 139, 627-639.

Murray-Rust, P., and Motherwell, S. (1978). Computer retrieval and analysis of molecular geometry. III. Geometry of the β-1'-aminofuranoside fragment. Acta. Crystallogr., Sect. B; Struct. Sci. B34, 2534-2546.

Murray-Rust, P., and Raftery, J. (1985). Computer analysis of molecular geometry. VI. Classification of differences in conformation. J. Mol. Graphics 3, 50-59.

Narayana, S. V. L. and Argos, P. (1984). Residue contacts in protein structures and implications for protein folding. Int. J. Peptide Protein Res. 24, 25-39.

Nyburg, S.C., and Faerman, C.H., (1985). A revision of van der Waals atomic radii for molecular crystals: N, O, F, S, Cl, Se, Br, and I bonded to carbon. Acta. Crystallogr., Sect. B; Struct. Sci. B41, 274-279.

Pauling, L. (1939). The Nature of the Chemical Bond. Cornell University Press, Ithaca.

Pimentel, G.C. and McClellan, A.L. (1960). The Hydrogen Bond. Freeman, San Francisco.

Rose, G. D., Geselowitz, A. R., Lesser, G. J., Lee, R., and Zehfus, M. H. (1985). Hydrophobicity of amino acid residues in globular proteins. Science 229, 834-838.

Rosenfield, R.E., Jr., Parthasarathy, R., and Dunitz, J.D. (1977). Directional preferences of nonbonded atomic contacts with divalent sulfur. 1. Electrophiles and nucleophiles. J. Am. Chem. Soc. 99, 4860-4862.

Rosenfield, R.E., Jr., Swanson, S.M., Meyer, E.F., Jr., Carrell, H.L., and Murray-Rust,P. (1984). Mapping the atomic environment of functional groups: turning 3D scatter plots into pseudo-density contours. J. Mol. Graphics 2, 43-46.

Schulz, G. E. and Schirmer, R. H. (1979). Principles of Protein Structure. Springer-Verlag, New York.

Singh, J., and Thornton, J. M. (1985). The interaction between phenylalanine rings in proteins. FEBS 191, 1-6.

Sutor, D.J. (1962). The C-H···O hydrogen bond in crystals. Nature 195, 68-69.

Taylor, R., and Kennard, O. (1982). Crystallographic evidence for the existence of C-H···O, C-H···N, and C-H···Cl hydrogen bonds. J. Am. Chem. Soc. 104, 5063-5070.

Taylor, R., Kennard, O., and Versichel, W. (1983). Geometry of the N-H···O=C hydrogen bond. 1. Lone-pair directionality. J. Am. Chem. Soc. 105, 5761-5766.

Thomas, K. A., Smith, G. M., Thomas, T. B., and Feldmann, R. J. (1982). Electronic distributions within protein phenylalanine aromatic rings are reflected by the three-dimensional oxygen atom environments. Proc. Natl. Acad. Sci. USA 79, 4843-4847.

Warme, P. K., and Morgan, R. S. (1978a). A survey of atomic interactions in 21 proteins. J. Mol. Biol. 118, 273-287.

Warme, P. K., and Morgan, R. S. (1978b). A survey of amino aid side-chain interactions in 21 proteins. J. Mol. Biol. 118, 289-304.

Aladdin: A Real Tool for Stucture-Based Drug Design

YVONNE C. MARTIN

EXPECTATIONS OF COMPUTATIONAL CHEMISTRY AND MOLECULAR GRAPHICS IN DRUG DESIGN

Molecular modeling groups in industry face a challenge different from those in academe. Specifically, we are generally outnumbered by at least 10:1 by synthetic organic chemists who are eager to have our input but who will move on without looking back if we are too slow to answer their questions. They expect computational chemists to either suggest for synthesis novel and patentable compounds that will possess the desired biological profile or to propose known compounds for testing in a new test. How do we meet these challenges?

Our experience has been that the actual design of new compounds from structure comes to an impass if the only means to suggest new compounds is a scientist sitting at a molecular graphics screen remembering images of previously seen structures and marrying these to concepts of required chemical groups. Frustrated by the meagre output of such molecular graphics-based design, we have written a computer program, ALADDIN, to aid in the design process [Van Drie, et al., 1989]. Specifically ALADDIN searches a database of three-dimensional structures to find those molecules that meet the specified substructural, geometric, and steric requirements. One can use ALADDIN to identify known compounds to test in a new assay. More exciting is its use to identify "template molecules" to which functional groups can be added to either

search for the proposed bioactive conformation or mimic a
known active molecule.

The task is made harder because usually today we do not know
the three-dimensional structure of the binding site on the
target biomolecule. Conceptually the work is divided into
the several stages elaborated in the following sections.
This report summarizes such a design process and the role
that ALADDIN might play in it.

STEP 1:   STUDY THE THREE-DIMENSIONAL
PROPERTIES OF KNOWN LIGANDS AND PROPOSE GEOMETRIC
AND CHEMICAL REQUIREMENTS FOR ACTIVITY

An analysis of the structure-activity relationships of the
known compounds that bind to the macromolecule is the first
step of the "receptor mapping" process. Since this field has
been well reviewed it is not necessary to detail the methods
[Marshall, et al., 1979; Martin and Danaher, 1988].
Essentially it involves comparing the three-dimensional
structures of active compounds to discover their common
three-dimensional features.

Receptor mapping includes identification of the atoms or
functional groups that appear to be necessary for affinity to
the target biomolecule as well as the geometric relationships
between them. Not only the low-energy conformation, but all
energetically accessible conformations must be compared.
Molecular graphics, conformational searching, and energy minimization are key computational aspects of this stage of the
investigation.

The surface that encloses all superimposed active compounds
defines the minimum size and shape of the macromolecular
binding site: regions in space occupied by inactive

molecules that possess the correct pharmacophore define putative regions occupied by the receptor.

In order to accomplish this task in a timely fashion we developed a database in which is stored the two-dimensional
structure of the molecules, their potencies in various biological tests, their calculated or measured physical properties, and the three-dimensional structures studied in our
molecular modeling and graphics [Martin, *et al.*, 1988]. This
is updated weekly with new two-dimensional structures and biological data from our MACCS/ORACLE databases. Literature
data of interest is entered manually. This database can be
searched by substructure or molecular similarly as well as
the value of or presence or absence of any data type.
MENTHOR is closely coupled to our molecular graphics system:
it is the storage medium for coordinates. This gives us the
ability to retrieve for display the proposed bioactive information of each of a set of molecules, etc.

Figure 1 shows our proposed model of the ligand binding site
for agonists of the dopamine D-2 receptor. Recently Cramer,
et.al. have shown that one can quantitate such notions
[1988]. Figure 2 shows contours of regions in space, if occupied, increase D-2 binding affinity [Seeman, *et al.*, 1988]
and those that decrease it [Lin and Martin, *in preparation*].
Thus methods exist to derive a computer-readable description
of the three-dimensional requirements that, if a molecule
possesses, it will show a particular biological property.

Of course, if all of the molecules tested are very conformationally flexible, then there might be no definitive answer
as to the bioactive conformation. In such a case one would
use ALADDIN to design compounds that probe this issue.
Specifically, one would search for different templates that
hold the functional groups in the relationships of each of
the possible bioactive conformations.

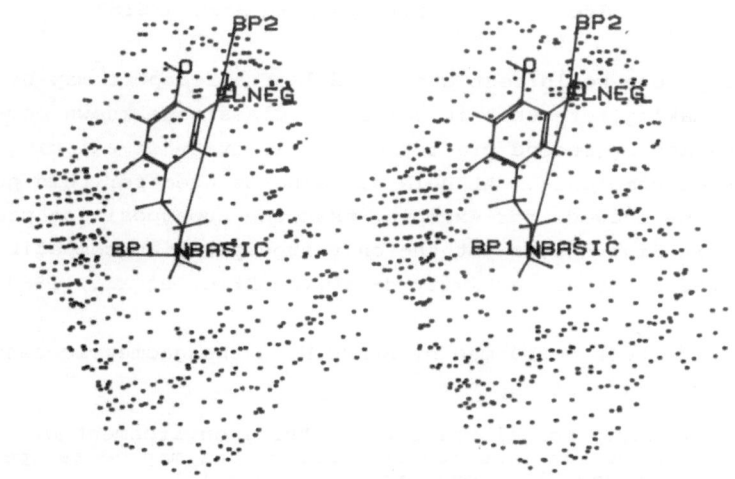

Figure 1. This figure shows dopamine enclosed by the union surface of active D-2 dopaminergic agonists. The more dense dots to the left of the figure represent the forbidden region, that is regions in space proposed to be required by the receptor. The requirements for D-2 dopaminergic activity are thus (1) a basic nitrogen atom at position NBASIC with a lone pair pointing toward BP1, (2) an electronegative atom at position ELNEG with a hydrogen-bonding hydrogen atom pointing toward position BP2, (3) the molecule must fit within the dotted surface.

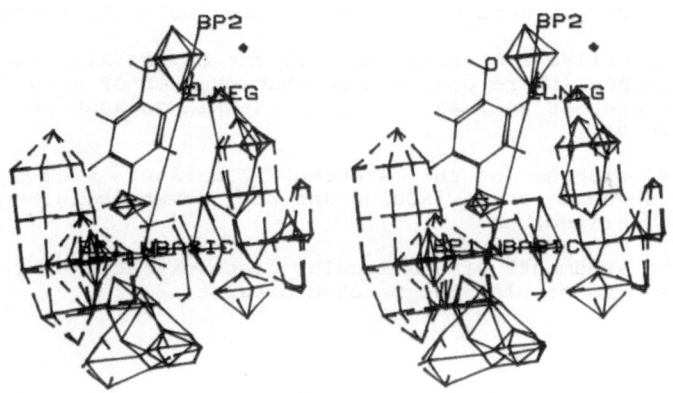

Figure 2. This figure shows dopamine and the proposed receptor binding points with the contours that define affinity. The dashed contours are regions in space, which occupied decrease affinity, whereas the solid contours are regions which increase affinity.

STEP 2: TEST THE PROPOSED HYPOTHESIS

Clearly, one would test one's model of a receptor map by
first making sure that it correctly classifies known compounds and by second predicting the activity of existing but
untested compounds. ALADDIN can also be used for this purpose. We showed, for example, that the superposition proposed in Figure 1 is not the only superposition compatible
with the structure-activity data [Van Drie, et al., 1989].

The first test performed by ALADDIN is the geometric test.
Input consists of:

> a description of the substructural environment of the
> atoms of interest (these descriptions may be as specific
> or as general as one wishes to make it);

> the geometric objects (points, lines, and planes) calcu-
> lated from the coordinates of the specified atoms; and

> the distance, angle, and torsion angle constraints that
> an XYZ dataset must match.

The output from ALADDIN consists of files that include the
following:

> the details of the calculation on each XYZ dataset,

> the 2-D structure and name of each molecule matched,

> a labelled 2-D structural diagram of the XYZ datasets
> matched (there can be more than one set of hits per XYZ
> dataset as well as more than one XYZ dataset per com-
> pound),

> 2-D diagrams of the matching XYZ datasets with the cor-
> responding atoms labeled and the geometric information
> summarized, and

> the arguments of a molecular graphics macro that can be
> used to display the matched datasets.

The steric test is then performed using as input a reference
molecule and atoms for superposition, the dot surface of the
binding site, and the molecular graphics macro arguments. It
checks each specified XYZ data set for atoms that are too

259

close to the surface to identify those molecules that fit within the union surface of the active molecules.

ALADDIN identified a number of molecules that, when tested, showed previously unknown dopamine D-2 agonist activity, two of which are shown in Figure 3.

STEP 3: DESIGN NOVEL COMPOUNDS AND/OR DISCOVER KNOWN COMPOUNDS WITH EXPECTED BIOACTIVITY

ALADDIN can also be used to design novel compounds that meet some geometric and substructural criteria. This is useful if one wishes to probe the bioactive conformation of a flexible ligand: sets of compounds would be designed to fix the key functional in geometric relationships characteristic of each of the low-energy conformations. Novel compound design is also the goal when one is certain of the geometric and steric requirements for bioactivity.

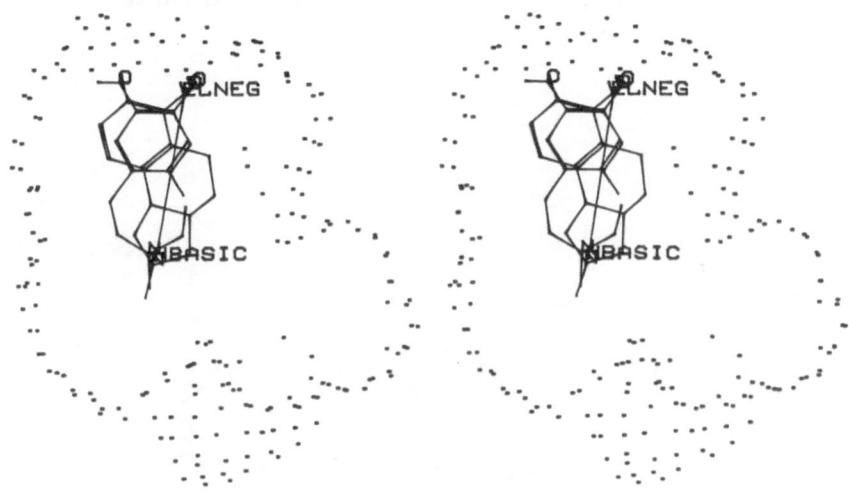

Figure 3. This figure shows two dopaminergic compounds identified by an ALADDIN search placed into the proposed D-2 binding site.

The following steps are used in such a design process:

> First an ALADDIN run identifies templates that have atoms in the appropriate geometric relationships but are of the wrong chemical nature.
>
> These templates are modified into the target compounds. This may be done using traditional molecular modeling and molecular graphics aided by the ALADDIN identification of the appropriate atoms. Alternatively, the two-dimensional connection tables of these templates can be changed into those of the desired molecules using a program called MODSMI and the modified connection tables used to generate three-dimensional coordinates with CONCORD [Rusinko, et al., 1988].
>
> The second ALADDIN run verifies that the designed molecules indeed meet the original criteria.

Figure 4 shows molecules designed by the latter process to match each of the three major conformations of dopamine.

DISCUSSION

While our work on ALADDIN was being conducted, others have also worked on the problem of three-dimensional substructure searching [Jakes, et al., 1987; Brint and Willett, 1987; and Sheridan, et al., 1989]. ALADDIN differs from these programs

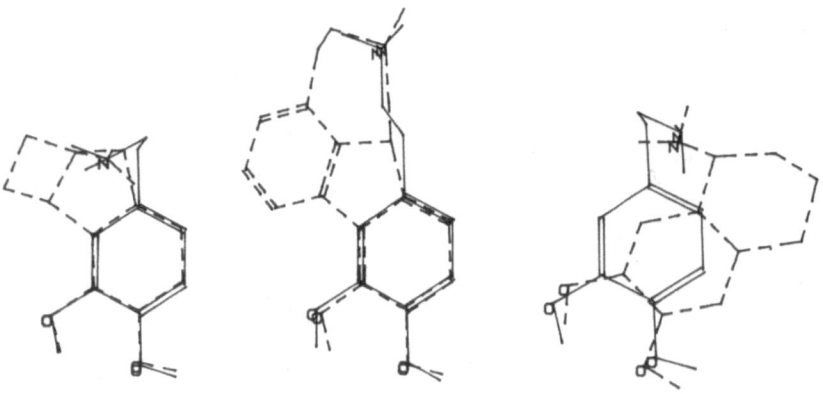

Figure 4. This figure shows in solid line the three major low-energy conformations of dopamine. Superimposed on each is a molecule designed by ALADDIN.

in it's closer integration with molecular modeling and molecular graphics programs, in its richer substructure and geometric object definition language, and in its use of a chemical information database so that criteria other than geometric or steric can be used in the query. There are deficiencies in the current implementation of ALADDIN that will be corrected. The most notable is that, some sort of screen will be used to limit the number of XYZ datasets that need undergo the detailed geometric test, as the other workers have done.

Others have addressed the problem of finding templates for automated structure design [Bartlett, *in press* and Lewis and Dean, 1989]. Again, the advantage of ALADDIN is its closer ties to molecular graphics and molecular modeling software. In addition, ALADDIN has a richer group of geometric constructs that can be used in the search process.

One limitation of all of these programs, including ALADDIN, is that they examine only the conformations already in the database [Martin, *et al., in press*]. One can accommodate conformational flexibility to some extent by a wise choice of atoms and constraints, for example to calculate the desired position of a hydroxyl hydrogen from the position of the oxygen and its attached carbon rather than use the position of the hydrogen atom in the database. However, this approach is not a general solution to the problem of conformational flexibility.

A further limitation of all of these programs is that they do not suggest molecules beyond the rules given to it. The user must perceive that a minor structural modification, such as enlargement of a ring or changing a single to a double bond, would result in another molecule that might fit.

If the objective of the design is to discover a ligand of a protein of known three-dimensional structure, then the program GRID [Goodford, 1984] might be used to identify sites at which it will be especially important to place atoms to satisfy electrostatic or hydrogen-bonding interactions. ALADDIN would be used to find the framework on which to place these groups in the correct geometric relationships. We have not fully tested the ability of ALADDIN for design to a known macromolecular binding site.

Many of the same concepts that are used in ALADDIN are also useful in the examination and comparison of the three-dimensional structures of proteins [Martin, *et al.*, *in press*].

In summary, the era of structure-based ligand design has been advanced by the availability of computer programs that use three-dimensional substructure searching techniques to recognize molecules that should be active in a particular biological test and to recognize templates from which other biologically active molecules can be constructed.

REFERENCES

Bartlett, P.A., G.T. Shea, S.J. Telfer, and S. Waterman. 1988. CAVEAT A program to facilitate the structure-derived design of biologically active molecules, in Chemical and

Biological Problems in Molecular Recognition. S. M. Roberts, S. V. Ley, and M. M. Campbell, eds, Roy. Soc. Chem. In Press.

Brint, A.T. and P. Willett. 1987. Pharmacophoric pattern matching in files of 3-D chemical structures: comparison of geometric searching algorithms. Journal of Molecular Graphics. 5:49-56.

Cramer III, R.D., D.E. Patterson, and J.D. Bunce. 1988. Comparative Molecular Field Analysis (CoMFA). 1. Effect of shape on binding of steroids to carrier proteins. Journal of the American Chemical Society. 110:5959-5967.

Goodford, P.J. 1985. A computational procedure for determining energetically favored binding sites on biologically important macromolecules. Journal of Medicinal Chemistry. 28:849-857.

Jakes, S.E., N. Watts, P. Willett, D. Bawden, and J.D. Fisher. 1987. Pharmacophoric pattern matching in files of three-dimensional chemical structures: evaluation of search performance. Journal of Molecular Graphics. 5:41-48.

Lewis, R.A. and P.M. Dean. 1989. Automated site-directed drug design: the concept of spacer skeletons for primary structure generation. Proc R. Soc. Lond. B. 236:125-140.

ewis, R.A. and P.M. Dean. 1989. Automated site-directed drug
esign; the formation of molecular templates in primary
tructure generation. Proc R. Soc. Lond. B. 236:141-162.

arshall, G.R., C.D. Barry, H.E. Bosshard, R.A. Dammkoehler,
nd D.A. Dunn. 1979. The conformational parameter in drug de-
ign: The active analog approach. In: "Computer-Assisted Drug
esign", (E.C. Olson and R.E. Christoffersen, eds.) American
hemical Society.Symposium 112, Washington. 205-226.

artin, Y.C., M.G. Bures, and P. Willett. 1989. Searching
atabases of three-dimensional structures. In: "Reviews in
omputational Chemistry". (K. Lipkowitz and D. Boyd, eds.) In
ress.

artin, Y.C. and E.B. Danaher. 1988. Molecular modeling or
eceptor-ligand interactions. In: "Receptor Pharmacology and
unction". (M. Williams., R. Glennon, and P. Timmermans,
ds.). Dekker, NY. 131-171.

artin, Y.C., E.A. Danaher, C.S. May, and D. Weininger. 1988.
ENTHOR, a database system for the storage and retrieval of
hree-dimensional molecular structures and associated data
earchable by substructural, biologic, physical, or geometric
roperties. Journal of Computer-Aided Molecular Design. 2:15-
9.

artin, Y.C. and T. Lin. unpublished observations.

DL, 1989: MACCS-II and MACCS-3D are software products from
olecular Design Limited, San Leandro, California, 94577.

usinko III, A., J.M. Skell, R. Balducci, C.M. McGarity, and
.S. Pearlman. 1988. CONCORD, A program for the rapid genera-
ion of high quality approximate 3-dimensional molecular
tructures, The University of Texas at Austin and Tripos
ssociates, St. Louis, Missouri.

eeman, P., M. Watanabe, D.Grigoriadis, J.L. Tedesco, S.R.
eorge, U. Svensson, J.L.G. Nilsson, and J.L. Neumeyer. 1985.
opamine D2 receptor binding sites for agonists. A tetrahe-
ral model. Molecular Pharmacology. 28:391-399.

heridan, R.P., A. Rusinko III, Nilakantan, and R.
enkataraghavan. 1989. Searching for pharmacophores in large
oordinate databases and its use in drug design. Proceedings
f the National Academy of Sciences (USA). In Press.

heridan, R.P. R. Nilakantan, A. Rusinko III, N. Bauman, K.
. Haraki, and R. Venkataraghavan. 1989. 3DSEARCH, A system
or three-dimensional substructure searching. Journal of
hemical Information and Computer Sciences. In Press.

an Drie, J.H., D. Weininger, and Y.C. Martin. 1989. ALADDIN:
n integrated tool for computer-assisted molecular design and
harmacophore recognition from geometric, steric, and sub-
tructure searching of three-dimensional molecular struc-
ures. Journal of Computer-Aided Molecular Design. In Press.

Index

1,2-dimethoxybenzene, 72
1,2,3-trimethoxybenzene, 72
2'deoxythymidylate, 1
2'deoxyuridylate, 1
4-amino-N-phenyl benzamides, 156, 157
4-carboxy analogs, 37
5-fluorodeoxuridine monophosphate, 1
5,10-methylenetetrahydrofolate, 1
10-propargyl-5,8-dideazafolate, 2, 5, 6, 7
19-norandrostenediol, 166, 167

Acceptor, 202
Active conformation, 162, 164, 169, 170, 171
Active site, 2, 3, 7, 47
Active site pocket, 35
Acylation, 33, 34, 39
Acylation reaction, 37
Acyl enzyme intermediate, 34
A-DNA, 124, 145
Adult respiratory distress syndrome, 30
ALADDIN, 254
    design, 259, 260
    input, 258
    output, 258
    prediction of activity, 259
    steric test, 258
Alanine dipeptide, 100, 101, 103
α-benzyl glutarimide, 152, 153, 154, 156, 159
α-helices, 85, 86, 88, 92
α-helix, 102

Alpha$_1$ proteinase inhibitor, 30
AMBER, 202, 203, 206
Amino acid side-chain, 175
Amino acid substitute, 96, 104
Anisole, 72
Antibacterial, 56
Anticonvulsant drugs, 151, 152, 153, 154, 156, 157, 159
Antimetabolites, 1
Antitumor drugs, 123, 124
Aromatic–aromatic interaction, 237
Aromatic–polar interactions, 244
Atomic charges, 155, 156
Autoimmune disease, 44
Azide, 34

Bacteriophage T4, 80
B-DNA, 123, 124, 126, 128, 130, 131, 135, 140, 141, 142, 143, 145, 147
Bent DNA, 124, 141
Benzyl clavulanate, 31, 32
β-conformation, 100
β-lactam, 30
    antibiotics, 30, 31, 37
    elastase inhibitors, 31, 32, 34, 35, 37
    inhibitor, 31
    ring, 34
β-lactamases, 30, 31, 33
β-lactams, 30, 35
β-turn, 108, 109
Bifurcated hydrogen bonds, 123, 129, 131, 140, 141, 142, 143

Binding affinity of drugs, 126
Binding conformation, 60
Bond length standards, 213
Brookhaven Protein Database, 231

Cambridge Structural Database, 201, 211, 231
  information content, 212
  software system, 212
Canyon, 11
Carbamazepine, 151, 156
Carbonyl, 244
Carboxyl analog of L-647,957, 38
Catalytic activity of mutants, 114, 117
Catalytic triad, 34
Cavity, 83
Cell attachment, 24
Cephalosporins, 37
Cephalosporin sulfone, 32, 36
Charge, 237
CHARMm, 38
Clavulanic acid, 31
Cluster analysis, 213
  Jarvis–Patrick, 218
  single-linkage, 218
  symmetry-modified, 218
Clustering, 245
CMTL, 115
CNDO method, 152
Co-minimization of molecules, 162, 168, 171
Computer-assisted, 201, 204
Computer graphics, 189
Conformation, 255, 256, 259, 260, 261
Conformational analyses, 67
Conformational analysis, 211
  cyclohexane, 215
  cyclopropyl-carbonyls, 213
  furanose sugars, 213
  primary esters, 213
  pseudorotation, 223
Conformational change, 3, 52, 53
Conformational distribution, 95, 96
Conformational equilibria, 96
Conformational restraints and constraints, 105
CONFOS, 164, 165, 166, 168

Cooperative-binding, 59
Cryogenic methods, 39
Crystallographic databases, 211, 229
Crystals, 81, 86
Crystal structures, 59
Cyclazocine, 169

Data base, 189, 190
Database, 201, 202, 256
Data base modelling, 189
Database searches, 254, 256, 261
Deoxythioguanosine, 44
Deoxyuridine monophosphate, 206
Design, 200, 201, 202, 203, 204, 205
Desolvation, 73
DHFR, 56, 58
Diaminopyrimidine, 69
Dideoxyinosine, 44
Difference map, 7
Diffusion, 39
Dihydrofolate reductase (DHFR), 1, 2, 3, 4, 5, 6, 8, 56
Dimethyl sulfoxide, 35
Diphenylhydantoin, 151, 156
Directed mutagenesis, 80
Dissimilarity coefficients, 218
Distamycin, 123, 124, 126, 127, 128, 129, 130, 131, 134, 135, 143, 145, 147
Disulfide bridges, 81, 89, 91, 92
Divalent sulfur, 233, 240
DNA, 200
DNA polymorphism, 124
DOCK, 201, 203, 205
Donor, 202
Dopamine D-2 receptor, 256, 259
Drug design, 161, 162

EETI, 115
Elastase, 30, 33
Elastase inhibitors, 37
Elastin, 29
Electron density, 194
Electron density maps, 175, 181
Electron diffraction, 165
Electron-distribution, 152, 156, 159
Electrophiles, 233, 240

Electrostatic, 116, 120, 202, 203, 204, 205
Electrostatic interactions, 66, 85
Emphysema, 29, 30
Enantiomer, 167
Enantiomeric specificity, 167
Energy contour map, 67, 69
Energy minimization, 52
Engineered protein, 84, 89, 92
Enkephalin, 170
Entropy, 86, 91
Enzyme, 56, 200, 201, 203, 204, 205, 206
Enzyme-inhibitor complex, 32, 39
Equilibrium constants, 106
Extra potential, 166, 169

Fentanyl, 172
FMDV loop, 15
Force fields, 162, 164, 165
Free energy barriers, 101
Free energy perturbation methods, 114,
    115, 117, 118, 119, 120
Free energy simulations, 96
FRODO, 189

Geometric and chemical requirements for
    activity, 255
Glycine dipeptide, 101, 102
GSTAT, 231

Halogen substituents, 152, 153, 154
Helix-dipole, 85
Helix→coil transition, 103
Hoechst 33258, 123, 124, 127, 128, 129,
    134, 144, 145, 148
Homology building, 185
Homology model building, 175
Hoogsteen geometry, 124, 129
Host-vs-graft response, 44
HRV-1A, 21
Human neutrophil, 32
Human neutrophil elastase, 29, 30, 31, 32,
    33, 37
Hydrogen bond, 2, 5, 6
Hydrogen bonding, 83, 84, 85, 86, 202,
    203, 204, 205

Hydrogen bonds, 50, 63, 230, 238
Hydrophilicity, 236
Hydrophobic, 49, 50, 64
Hydrophobic effect, 81
Hydrophobic effects, 24, 50
Hydrophobic interactions, 234
Hydrophobicity, 234, 237
Hydrophobic pocket, 6
Hydroxyl amine, 34

ICAM-1, 9
In vitro data, 163
In vivo data, 163, 164
Inactive forbidden regions, 255
In-crystal minimization, 164
Industry, 254
Inhibitor, 58
Interaction energy, 74
Interactive computer graphics, 175
Intercalator, 124, 135, 148
Intermolecular non-bonded contacts, 231
Intramolecular, 231

L-647,256, 35, 36
L-647,957, 32, 33, 35, 36, 37, 38
Laue method, 39
Leucine enkephalin, 170
Leukemia, 44
Ligand, 204, 205
Low energy conformation, 164
Lysozyme, 80, 81, 83, 84, 85, 86, 87, 88,
    89, 91, 92

Maximal electroshock (MES), 151, 152,
    153, 154, 156, 159
Mechanism of action, 39
MENTHOR, 256
Methotrexate, 3, 5, 63
Methoxy groups, 57
Methylenetetrahydrofolate, 206
Microenvironment, 235
Microwave spectroscopy, 165
MIDAS, 202
Minor groove binding drugs, 123, 147
MNDO, 155

Model, 162, 167, 171, 172, 173, 174
Model building, 221, 226
Modeling, 206
Model regularisation, 193
Molecular association, 230
Molecular dynamics, 74, 95, 104, 105,
   114, 115, 116, 117
Molecular graphics, 166, 167, 169,
   173
Molecular mechanics, 67, 157, 158, 161,
   162, 163, 164, 165, 166, 168, 169,
   174, 203, 206
Molecular recognition, 230
Molecular surface, 201
MOLEDIT, 36
MOPAC, 155
Morphine, 168, 169
MULTAN, 36
Multivariate statistics, 213
Mutations, 21, 92

NADPH, 56
Netropsin, 123, 124, 126, 127, 129, 130,
   131, 134, 142, 143, 144, 145, 147,
   148
Neutron diffraction, 165
Neutrophil, 29
NMR structural refinement, 114
Non-bonded interactions, 230
Nucleophiles, 233, 240
Nucleophilic attack, 34, 35
Nucleoside, 43

O, 189
O data base, 190
O-menus, 191
   _Bones, 192
   _Draw_Mol, 191
   _Lsq_align, 193
   _Manip, 193
   _Map, 191
   _Paint_Mol, 194
   _Proleg, 193
   _Refi, 193
   _RSR, 192
   _Sketch_Mol, 194
Opiate μ receptor, 172

ORTEP, 36
Ortho position, 248

Pattern recognition, 211
Penicillanic acid sulfone, 31
Penicillopepsin, 203, 204, 205
Peptide chloromethyl ketone, 32
Peptide conformation, 100
Pharmacological data, 163, 171
Pharmacology, 164, 173
Phenyl, 244
Phenylalanine, 237
Phenylpiperidinopyridazines, 153, 154,
   155, 156
Phosphate binding site, 47
PNP, 44, 45, 46, 47, 49, 51, 52, 53, 55
Porcine pancreatic elastase, 31, 32, 33, 36,
   37, 38
Porcine pancreatic elastase inhibitor com-
   plexes, 33, 35
Porcine pancreatic elastase L-647,957
   complex, 33
Potential, 202, 203, 204, 205
Potential energy hypersurface, 240
Predicting activity, 257
Preferential association, 233, 235
Principal component analysis, 213
Proline, 86
Propeller twists, 123, 141, 142, 148
Protein, 200, 202, 203, 204
Protein design, 114
Protein engineering, 175, 176
Protein-ligand interactions, 206
Protein structure, 175
Pseudo-density, 245
Puckering parameters, 223
Purine, 50
Purine and phosphate binding site, 50
Purine binding site, 47
Purine nucleoside phosphorylase, 43

QUANTA, 38
Quinazoline, 3, 6, 7

Random coil, 106
Real space refinement, 192

Receptor, 161, 162, 163, 169, 171, 172, 173, 174
Receptor mapping, 255
Receptors, 19, 200, 201, 202, 203, 204, 205
Recombinant, 45
Rheumatoid arthritis, 30
Rhinovirus, 9, 204
Ribose, 48
Rotamer conformations, 175
Rotamers, 181

Selective affinity, 66
Selective binding, 67
Serine hydroxymethyltransferase, 1
Serine protease, 29
Serotypes, 9, 20
Shared surface area, 234
Shared surface correlation, 235
Site-specific mutagenesis, 24, 114, 115
Slow substrates, 39
Solvation, 72
Solvation energy, 71, 73
sp^2 hybridization geometry, 232
sp^3 hybridization geometry, 232
Stability, 96, 105, 106
Stability mutants, 110
Steric, 201
Strain, 83, 87, 88, 91
Structural analysis, 176
Structural motifs, 175
Structure/activity information, 173
Structure-activity relationships, 255
Structure analysis, 175
Subcutaneous Metrazol (scMET), 151
Substilisin, 119
Substructure, 231
Sulbactam, 31
Sulbactam benzyl ester, 32
Sulphur, 238
Sulphur lone pairs, 240

Superimposed active compounds, 255
Surface area, 234
Synchrotron radiation, 46
Synchrotron X-ray source, 39
Synthesis, 254

T4-lysozyme, 110, 120, 121
T-cell, 44, 50
Tetrahydrofolate, 56
Thermal stability, 114, 115
Thermodynamic cycle, 105
Thermolysin, 118
Thermostability, 80, 81, 83, 86, 87, 92
Three-dimensional, 200
Thymidylate synthase (TS), 1, 2, 3, 5, 6, 7, 8, 206
Topological symmetry, 215, 221
Trimethoprim, 5, 56
Trimethoxyphenyl, 60
Triose phosphate isomerase, 116, 117

U-50,488, 169
Uncoat, 12
Union surface, 255

van der Waals, 120
van der Waals interactions, 66
van der Waals radii, 241
Vapor pressures, 73

Water solubilities, 73
WIN compounds, 12, 14

X-ray crystallography, 32, 58
X-ray diffraction, 123, 126, 165

Z-DNA, 124